# THERMAL ENERGY
# AT THE NANOSCALE

# Lessons from Nanoscience: A Lecture Note Series

ISSN: 2301-3354

**Series Editors:**   Mark Lundstrom and Supriyo Datta
*(Purdue University, USA)*

"Lessons from Nanoscience" aims to present new viewpoints that help understand, integrate, and apply recent developments in nanoscience while also using them to re-think old and familiar subjects. Some of these viewpoints may not yet be in final form, but we hope this series will provide a forum for them to evolve and develop into the textbooks of tomorrow that train and guide our students and young researchers as they turn nanoscience into nanotechnology. To help communicate across disciplines, the series aims to be accessible to anyone with a bachelor's degree in science or engineering.

More information on the series as well as additional resources for each volume can be found at: http://nanohub.org/topics/LessonsfromNanoscience

*Published:*

Vol. 1   Lessons from Nanoelectronics: A New Perspective on Transport
*by Supriyo Datta*

Vol. 2   Near-Equilibrium Transport: Fundamentals and Applications
*by Mark Lundstrom, Changwook Jeong and Raseong Kim*

Vol. 3   Thermal Energy at the Nanoscale
*by Timothy S Fisher*

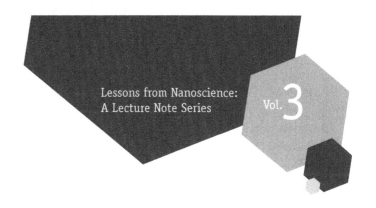

Lessons from Nanoscience:
A Lecture Note Series

Vol. 3

# THERMAL ENERGY AT THE NANOSCALE

## Timothy S Fisher

*Purdue University, USA*

**World Scientific**

NEW JERSEY • LONDON • SINGAPORE • BEIJING • SHANGHAI • HONG KONG • TAIPEI • CHENNAI

*Published by*

World Scientific Publishing Co. Pte. Ltd.

5 Toh Tuck Link, Singapore 596224

*USA office:* 27 Warren Street, Suite 401-402, Hackensack, NJ 07601

*UK office:* 57 Shelton Street, Covent Garden, London WC2H 9HE

**Library of Congress Cataloging-in-Publication Data**
Fisher, Timothy S., 1969–
    Thermal energy at the nanoscale / Timothy S. Fisher (Purdue University, USA).
        pages cm. -- (Lessons from nanoscience ; vol. 3)
    Includes bibliographical references and index.
    ISBN 978-9814449779 (hardcover : alk. paper) --
    ISBN 978-9814449786 (softcover : alk. paper)
    1. Nanostructured materials--Thermal properties. 2. Nanostructured materials--Transport
properties. 3. Energy storage. 4. Thermal conductivity. 5. Heat--Transmission. 6. Transport
theory. I. Title.
    QC176.8.T4F57 2013
    620.1'1596--dc23
                                                            2013028627

**British Library Cataloguing-in-Publication Data**
A catalogue record for this book is available from the British Library.

Typeset by Stallion Press
Email: enquiries@stallionpress.com

Printed in Singapore

To my wonderful and amazing wife Amy

# Preface

These notes provide a detailed treatment of the thermal energy storage and transport by conduction in natural and fabricated structures. Thermal energy by two main carriers–phonons and electrons–are explored from basic principles. For solid-state transport, a common Landauer framework is used for heat flow, and issues including the quantum of thermal conductance, ballistic interface resistance, and carrier scattering are elucidated. Bulk material properties, such as thermal conductivity, are derived from transport theories, and the effects of spatial confinement on these properties are established.

The foregoing topics themselves are not unique as elements in a book; many other outstanding texts cover these topics admirably and are cited in context herein. At the same time, the present content emphasizes a basic theoretical framework based on the Landauer formalism that is as self-consistent as possible, not only internally but also with respect to similar efforts in this book series on the subject of electrical transport. The other series titles, written by Profs. Supriyo Datta and Mark Lundstrom, have therefore provided much inspiration to the present work, as have my related conversations with these two amazing colleagues. The end result is (hopefully) an accessible exposition on the foundations of the subject that remains concise by avoiding lengthy digressions into the vast array of related contemporary research topics. At the same time, it is my hope that readers, after studying this work, will be ready to enter the field well-equipped to contribute to this wonderful body of research and community of researchers.

*T. S. Fisher*

# Acknowledgments

This text has been thoroughly inspired by the large number of outstanding students whom I have been privileged to teach both in the classroom and laboratory during my career. The content of this text has been refined over the years through teaching students at Vanderbilt University, Purdue University, and the Jawaharlal Nehru Centre for Advanced Scientific Research (Bangalore, India), as well as those from around the world who participated in the first offering of an online course by the same name, first delivered through the nanoHUB-U initiative in Spring 2013. I convey particular gratitude to students Alfredo Tuesta, Anurag Kumar, Guoping Xiong, Jeff Engerer, Kim Saviers, Menglong Hao, and Stephen Hodson for assistance with proofreading and indexing. The nanoHUB-U team, and particularly Amanda Buckles, Joe Cychosz, Erich Huebner, and Mike McLennan, provided tremendous support in launching the class and allowing me to focus on content, most of which appears herein. I also express appreciation to members of Purdue's Mechanical Engineering Heat Transfer faculty, a group with whom I am humbly privileged to serve; and particularly in the context of this book, I express gratitude to Professors Jayathi Murthy (now at UT-Austin), Xiulin Ruan, and Xianfan Xu, each of whom has inspired substantial content herein. Other Purdue faculty colleagues whose influence has significantly shaped my interpretation of the subject matter include Supriyo Datta, Bob Lucht, Mark Lundstrom, Ron Reifenberger, Tim Sands, and Ali Shakouri. The content herein draws from many sponsored research projects in which I have participated over the years, and I convey my sincere appreciation to those sponsors. In terms of active research projects during the writing of the book, the most relevant is that from the US Office of Naval Research (Program Manager: Dr. Mark Spector) on interfacial heat transfer. I also thank the publisher, World Scientific, and

particularly Song Yu for supporting this book series and assisting with the
publication details.

Lastly, I offer my most sincere thanks and recognition to Sridhar Sada-
sivam and Ishan Srivastava, two doctoral graduate students in my group at
Purdue. Sridhar has served impeccably as a sounding board for the explanations and content in the text, as well as providing great help in compos-
ing and organizing graded content for the companion online course offered
through nanoHUB-U. Ishan has patiently tolerated my pedestrian capabil-
ities in graphic arts and created most of the graphics contained herein. He
has also developed a suite of simple, web-accessible simulation tools (using
Wolfram's CDF driver) for use in the online course that draws from the
content here. Aside from the foregoing specific contributions, our regular meetings to discuss ideas, explanations, theory, and content for these
notes and the online course have been tremendously invigorating. In these
days of much chaos for academic researchers, with the various and sundry
demands of our profession, finding time to focus on what really matters
with these two gifted colleagues has been delightful; I thank them for their
engagement.

# Contents

# Nomenclature

| | |
|---|---|
| $\alpha$ | thermal diffusivity (length$^2$/time) |
| $\beta$ | inverse of thermal energy, $(k_B T)^{-1}$ (1/energy) |
| $\chi$ | carrier energy scaled by $k_B T$ (-) |
| $\eta_a$ | unit cells per volume of real space (1/volume) |
| $\eta_e$ | volumetric electron density (1/volume) |
| $\hat{G}_Q$ | quantum of thermal conductance (energy/time/temperature) |
| $\kappa$ | thermal conductivity (power $\times$ length/('area' $\times$ temperature)) |
| $\Lambda$ | particle mean free path (length) |
| $\mathcal{D}$ | plate bending stiffness (force $\times$ distance $=$ energy) |
| $\mathcal{F}$ | plate loading (force/area) |
| $\mathcal{L}$ | boundary scattering length scale (length) |
| $\mathcal{T}$ | carrier transmission function (-) |
| $\mu$ | mass density of a continuum string (mass/length) |
| $\nu$ | Poisson ratio (-) |
| $\Omega$ | number of possible states of a statistical ensemble (-) |
| $\omega$ | frequency (radians/time) |
| $\omega_D$ | Debye frequency (radians/time) |
| $\omega_E$ | Einstein frequency (radians/time) |
| $\phi$ | emitter work function (energy) |
| $\rho$ | mass density (mass/volume) |
| $\sigma$ | scattering cross section (area) |
| $\sigma_e$ | electrical conductivity (current/(length $\times$ voltage)) |
| $\tau$ | scattering time (time) |
| $\tau^{-1}$ | scattering rate (1/time) |
| $\tau_b^{-1}$ | boundary scattering rate (1/time) |
| $\theta_D$ | Debye temperature (temperature) |
| $\theta_E$ | Einstein temperature (temperature) |

$\theta_F$      Fermi temperature (temperature)

$\tilde{G}'_Q$      scaled spectral thermal conductance (-)

$\varepsilon$      boson energy (energy)

$\vec{a}_i$      real-space lattice translational vectors (length)

$\vec{b}_i$      reciprocal lattice translation vectors (lattice)

$\vec{G}$      reciprocal lattice vector (1/length)

$\vec{R}$      real-space lattice vector (length)

$\vec{v}_g$      group velocity (length/time)

$a$      lattice constant (length)

$c$      phase velocity (length/time)

$c_0$      speed of light in vacuum, $2.99792458 \times 10^8$ m/s

$c_v$      volumetric specific heat (energy/(volume × temperature))

$D(\omega)$      density of boson states, frequency basis (time/volume)

$D(\varepsilon)$      density of boson states, energy basis (volume energy)$^{-1}$

$D(E)$      density of fermion states, energy basis (volume energy)$^{-1}$

$D(K)$      density of boson states, **k**-space basis (length/volume)

$D^\beta_\alpha$      dynamical matrix (force/(length × mass))

$E$      energy (energy)

$E_b$      bond energy (energy)

$E_F$      Fermi energy (energy)

$E_Y$      Young's modulus (force/area)

$E_{\text{vac}}$      vacuum energy level (energy)

$F$      boundary scattering fitting factor (-)

$F$      force on an atom due to bond stretching (force)

$f^o_i$      equilibrium carrier distribution function (-)

$f(t)$      forward-wave string displacement (length)

$g$      spring constant of an interatomic bond (force/length)

$G'_Q$      spectral thermal conductance (power/temperature, per unit frequency for phonons, or per unit energy for electrons)

$g(t)$      reflected-wave string displacement (length)

$G_Q$      thermal conductance (power/temperature)

$h$      plate thickness (length)

$J$      electrical current density (current/'area')

$J_Q$      heat flux (power/'area')

$K$      phonon wavevector (1/length)

$k$      electron wavevector (1/length)

$K_D$      Debye wavevector (1/length)

$k_F$      Fermi wavevector (1/length)

| | |
|---|---|
| $L_e$ | Lorenz number, dimensionless constant $\times \left(\frac{k_B}{q}\right)^2$ |
| $m$ | atomic mass (mass) |
| $M(\omega)$ | number of phonon modes (-) |
| $M(E)$ | number of electron modes (-) |
| $m_e$ | electron mass, $9.10938188 \times 10^{-31}$ kg |
| $M_{dD}(\omega)$ | phonon mode density, $d =$ system dimension (1/'area') |
| $M_{dD}(E)$ | electron mode density, $d =$ system dimension (1/'area') |
| $N$ | number of atoms (-) |
| $N'$ | electron number (-) |
| $N_A$ | Avogadro's number, $6.0221415 \times 10^{23}$ (-) |
| $n_i$ | defect density of impurity scatterers (1/volume) |
| $N_{dD}$ | number of allowed phonon states, $d =$ system dimension (-) |
| $N_k$ | number of allowed electron states (-) |
| $N_K$ | number of phonons with wave vector $K$ (-) |
| $P$ | acoustic wave power (energy/time) |
| $P_\nu$ | probability of a statistical state (-) |
| $q$ | elementary electron charge, $1.602 \times 10^{-19}$ C |
| $r$ | distance coordinate (length) |
| $R_b$ | thermal boundary (interface) resistance (temperature/power) |
| $R_b''$ | area-normalized thermal boundary (interface) resistance (area×temperature/power) |
| $S$ | entropy (power/temperature) |
| $t_{12}$ | interfacial energy transmittance from medium 1 to medium 2 (-) |
| $U$ | internal energy (energy) |
| $U$ | potential energy (energy) |
| $u$ | atomic displacement away from equilibrium (length) |
| $u$ | specific internal energy (energy/volume) |
| $u'(x)$ | spectral energy density (energy/volume, per unit $x$, where $x$ is a spectral quantity such as frequency or wavelength) |
| $v_a$ | acoustic wave velocity (length/time) |
| $v_F$ | Fermi velocity (length/time) |
| $y(x,t)$ | total string displacement (length) |
| $Z$ | acoustic impedance of a string under tension (mass/time) |

# List of Figures

# List of Tables

Chapter 1

# Lattice Structure, Phonons, and Electrons

## 1.1 Introduction

Guessing the technical background of students in a course or readers of a book is always a hazardous enterprise for an instructor, yet one must start a book or a course *somewhere* on the landscape of knowledge. Here, we begin with some essential concepts from condensed-matter physics and statistical mechanics. The definition of essential, too, is questionable and is presently intended to be information that recurs too frequently in the later parts of the text to leave the requisite information to the many excellent reference sources on these subjects.

Our overarching objective is to develop the tools required to predict thermal transport in structures such as the one shown in Fig. 1.1. Arguably the most important thermal characteristic of an object is its thermal conductivity ($\kappa$) defined as:

$$\kappa \equiv \frac{[\text{rate of heat flow (in W)}] \times [\text{object length (in m)}]}{[\text{cross-sectional area (in m}^2)] \times [\text{temperature drop (in K)}]}. \quad (1.1)$$

For roughly a century, thermal conductivity was considered a basic material property in the engineering sense (e.g., with minor accommodation for variations in temperature), and therefore, the effects of the geometric terms in Eq. (1.1) were assumed to normalize with the others such that the final property was independent of size and shape. However, with the advent of microscale fabrication (and later nanoscale fabrication), the technical community was able to create tiny materials that exhibited deviations from the size-independent property assumption. In such circumstances, knowledge of not only a material's size and shape becomes crucial but also the details of the atomic-scale carriers of thermal energy (Chen, 2005). At this level, in order to retain the utility of the concept of thermal conductivity (and it does remain useful for many purposes) we need to understand many additional factors, including:

- What type of quantum-mechanical carrier dominates heat flow in the material?
- How is thermal energy distributed among these carriers?
- How fast do the carriers move through the material?
- How much thermal energy does each carrier hold as it moves?
- How do the carriers scatter as they move through the material?
- How do the boundaries and interfaces impede carriers?

The answers to these questions require a much deeper perspective on the mechanisms of thermal energy transport than is provided in traditional engineering expositions on heat conduction. Thus we embark here on the first of two background chapters: the present on lattice structure and the subsequent on statistics of energy carriers.

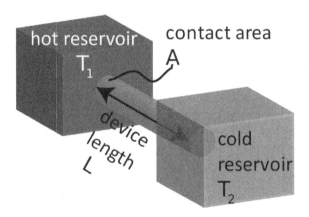

Fig. 1.1 Schematic of a general contact-device-contact arrangement.

The study of thermal energy in any material should rightly begin with a description of the material itself, for thermal energy, unlike other forms of energy such as optical, electronic, and magnetic, is routinely generated, stored, and transported by a diverse set of 'carriers'. The reason for broader context of thermal energy derives from the second law of thermodynamics, which dictates that all forms of energy tend toward disorder (or 'thermalization'). In this text, we will make every reasonable attempt to unify the analysis, i.e., to generalize concepts so that they apply to multiple carriers, but this objective is occasionally elusive. In such cases, the text will make clear the relevant restrictions by carrier and material types. The list of interesting materials and physical structures is almost endless, and therefore given the subject of 'nanoscale' physics, the text begins with an admittedly cursory treatment of interatomic bonding but then highlights where possible a compelling structure — the graphene carbon lattice — to illustrate important and unique thermal behavior at the nanoscale.

## 1.2 Atom-to-Atom Bonding in Solid Lattices

The details of interatomic bonding determine a broad assortment of physical material properties, ranging from mechanical strength to electrical conductivity. The primary interest here relates to the resultant vibrational characteristics of atoms that exist in an ordered arrangement, i.e., in a regular crystal. However, we start with a simpler situation: that of a diatomic molecule.

Figure 1.2 shows a schematic of two atoms separated by an equilibrium distance $r = r_0$ about which the atoms vibrate at various (but restricted) frequencies. A generic potential energy field $U(r)$ between the atoms is shown in the bottom half of the figure, revealing the strong repulsive force $(F = -\partial U/\partial r)$ when the atoms are close together $(r < r_0)$. The minimum energy (at $r = r_0$) corresponds to the bond energy, as the potential energy asymptotes to zero when the atoms are pulled apart $(r \to \infty)$.

The mathematical form of the potential can be very complicated and is itself the subject of intensive research through both first-principles (*ab initio*) approaches such as density functional theory (Saha *et al.*, 2008) and empirically derived potentials (Tersoff, 1988). For the time being, we consider a simplification of the potential, focusing on the near-minimum region where the potential is typically well approximated by a parabolic relation with respect to the equilibrium displacement $u = r - r_0$ such that $U \sim u^2$. The constant of proportionality plays an important role in the dynamics of molecules and lattices, for it contains the *effective spring constant* $g$ of the bond:

$$U = \frac{1}{2}gu^2. \tag{1.2}$$

This so-called harmonic approximation is depicted in Fig. 1.3. We note that lattice vibrations typically involve small displacements; therefore, the

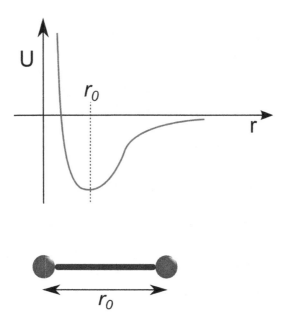

Fig. 1.2    Variation of potential energy field $U(r)$ with interatomic distance $r$. $r = r_0$ corresponds to the equilibrium separation with minimum potential energy.

harmonic approximation tends to predict the overall vibrational states (or what we will call the *vibrational eigenspectrum*) with good accuracy. The deviations, or anharmonicities, however, play an important role in phonon scattering, as discussed in Chapter 5.

One issue that we will cover only briefly is how such bonds form. Referring to Fig. 1.4, when two self-contained atoms [Fig. 1.4(a)] are brought together [Fig. 1.4(b)], their electrons can interact and begin to share orbitals. However, the energies of the orbitals must change because of restrictions imposed by the Pauli exclusion principle on the quantum states of electrons; therefore, upon bonding, the energy levels depicted by horizontal lines in Fig. 1.4, undergo small shifts.

These electronic interactions define the nature and strength of interatomic bonds and can produce many different bond types and energies ($E_b$), including:

- van der Waals: weak bond due to dipole moments, $E_b \sim 0.01$ eV
- Hydrogen: due to electronegative atoms (e.g., O in $H_2O$), $E_b \sim 0.1$ eV

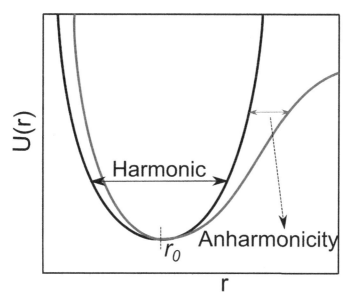

Fig. 1.3 Harmonic approximation to the real interatomic potential with anharmonicity. At small displacements, the harmonic potential is a good approximation.

- Covalent: atoms share valence electrons (e.g., Si and diamond), $E_b = 1 \sim 10$ eV
- Ionic: one atom gives up its electron, forms ions with Coulombic binding forces, $E_b = 1 \sim 10$ eV
- Metallic: like covalent bonds, but with freely moving electrons, $E_b = 1 \sim 10$ eV

We will focus on thermal energy in solid materials, but some of the content such as kinetic theory in Chapter 3 applies equally well to fluid phases. Within the array of solid-state materials, single-crystal structures are the most amenable for initial study, although even these structures become rather complex in three dimensions with various atomic arrangements such as face-centered cubic (fcc), body-centered cubic (bcc), and diamond configurations that are perhaps most familiar to readers. To minimize digression, here we refer the reader to the many excellent textbooks on solid-state physics (Ashcroft and Mermin, 1976; Kittel, 2007) and crystallography (De Graef and McHenry, 2012) for advanced treatment of 3D crystals.

We will focus on one- and two-dimensional lattices for the sake of expediency and because the 2D graphene lattice has high contemporary

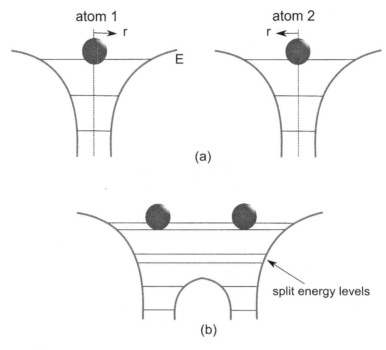

Fig. 1.4   (a) Two isolated, self-contained atoms and associated electron energy states. (b) Quantized energy states upon bond formation between the two isolated atoms. Energy levels are modified as electron orbitals become shared in a bond.

scientific and technological importance. A simple 1D structure is obtained by repeating the diatomic arrangement of Fig. 1.2 indefinitely. Figure 1.5 shows the resulting configuration, with each atom of mass $m$ connected to its neighbor by a bond with spring constant $g$. The equilibrium separation between atoms is represented by the lattice constant $a$. Somewhat surprisingly, this simple, idealized structure will enable us to develop almost all the essential tools for analysis of lattice vibrations and their quantum manifestation—called phonons.

Because an ideal crystal extends infinitely in all directions, we must find a way to concentrate the analysis on a smaller region. Fortunately, the regular order, or periodicity, of a crystal lattice makes this task straightforward. A *primitive unit cell* of a lattice is one that, if repeated throughout all space by well-defined translational vectors, would fill the space entirely and with no overlapping regions or void spaces. Figure 1.6 shows an example for a 2D monatomic rectangular lattice. Several possible shapes, positions, and

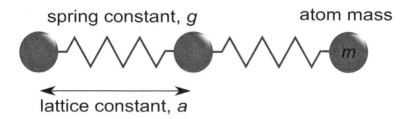

Fig. 1.5   An ideal 1D crystal modeled as periodic atom-spring-atom system.

orientations of the primitive unit cell exist for this lattice, as indicated by the shaded regions. The arrows denote *basis vectors* $(\vec{a}_i)$ that define the periodic translation of the unit cells throughout the domain. The set of all possible translations by integer indexing of basis vectors forms a so-called Bravais lattice, whose discrete points are given by the lattice vector $\vec{R}$:

$$\vec{R} = \sum_i n_i \vec{a}_i \overset{\text{in 2D}}{=} n_1 \vec{a}_1 + n_2 \vec{a}_2. \tag{1.3}$$

For the linear 1D chain, the sole lattice vector is simply the lattice constant $a$.

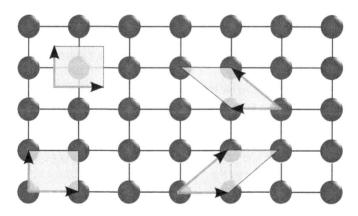

Fig. 1.6   An ideal 2D monatomic rectangular lattice represented by periodic translation of valid primitive cells (shaded green) defined by green basis vectors.

As might be expected given the complexity of our natural world, a Bravais lattice alone cannot describe the atomic positions of all real crystals. For such cases, we resort to defining the positions of multiple atoms (usually two) at each nodal site in the Bravais lattice. This approach is quite understandable for compounds such as crystalline $SiO_2$ (quartz), but it is also necessary to describe the lattice geometry of some monatomic crystals, including technologically important ones such as silicon and diamond. Figure 1.7 shows the crystal construction in 2D, with the multi-atom basis pair placed regularly on spatially distributed Bravais lattice points.

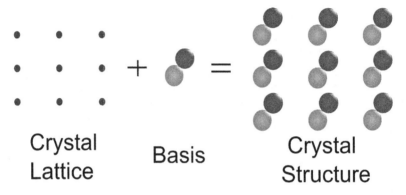

**Crystal Lattice      Basis      Crystal Structure**

Fig. 1.7    Structure of a two-atom basis crystal. Each nodal site of the lattice contains a two-atom basis that defines the complete crystal structure upon translation through all possible lattice vectors.

One of the most fascinating 2D lattices, and one of intense contemporary study, is graphene, which consists entirely of carbon atoms in hexagonal arrangement on a 2D plane as shown in Fig. 1.8. The equilibrium distance between nearest carbon atoms is $\tilde{a} = 1.42$ Å, where the '~' denotes a bond length (often the lattice constant and bond lengths differ for more complex lattices). Different edge configurations are possible in graphene, and the two most common are shown in the figure. Importantly, graphene is one of the monatomic structures that requires the addition of a basis atom to describe the full lattice. Its basis vectors, as shown in Fig. 1.9, are:

$$\vec{a}_1 = \frac{3}{2}\tilde{a}\hat{x} + \frac{\sqrt{3}}{2}\tilde{a}\hat{y}$$
$$\vec{a}_2 = \frac{3}{2}\tilde{a}\hat{x} - \frac{\sqrt{3}}{2}\tilde{a}\hat{y}. \tag{1.4}$$

The vector that connects the primary and basis atoms within a unit cell is simply $\vec{a}_b = \tilde{a}\hat{x}$.

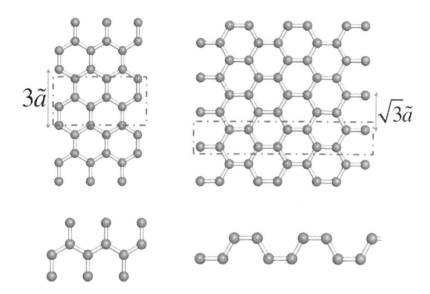

Fig. 1.8  Graphene nanoribbon crystal structure.  The left structure is the armchair configuration; the right structure is zigzag.  Dashed rectangles represent a graphene nanoribbon unit cell.  The unit cell for each configuration is displayed below the crystal lattice structure.

## 1.3   Mathematical Description of the Lattice

The analysis of crystals can seem challenging in comparison to that of individual molecules because of the former's vast size.  To overcome this challenge, we take advantage of a crystal lattice's *translational symmetry*. This approach requires a mathematical description that inverts space such that large entities become small.

We describe something large in terms of small things in the common way–with Fourier transforms.  We start again with 1D chain of atoms and allow for the possibility that these atoms have a distributed mass density $\rho$.  Perfect periodicity with lattice constant $a$ (see Fig. 1.10) implies that:

$$\rho(x + ma) = \rho(x), \tag{1.5}$$

where $m$ is any integer.

$$\tilde{a} = 1.42\,\text{Å}$$

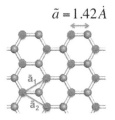

Translation vectors:

$$\vec{a}_1 = \tfrac{3}{2}\tilde{a}\hat{x} + \tfrac{\sqrt{3}}{2}\tilde{a}\hat{y}$$

$$\vec{a}_2 = \tfrac{3}{2}\tilde{a}\hat{x} - \tfrac{\sqrt{3}}{2}\tilde{a}\hat{y}$$

Fig. 1.9    Basis vectors for the graphene crystal structure.

Fig. 1.10    Perfect periodic 1D chain of atoms with lattice constant $a$.

Each density function $\rho(x)$ and $\rho(x+ma)$ can be expanded in a Fourier series such that Eq. (1.5) becomes:

$$\rho(x) = \sum_n \rho_n \exp\{iG_n x\}$$

$$= \rho(x+ma) = \sum_n \rho_n \exp\{iG_n(x+ma)\}$$

$$= \sum_n \rho_n \exp\{iG_n x\}\exp\{iG_n ma\}, \qquad (1.6)$$

$$\to \exp\{iG_n ma\} = 1 \to G_n ma = 2\pi \times \text{integer}, \qquad (1.7)$$

where $n$ and $m$ are indexing integers. The last relation, Eq. (1.7), severely restricts the possible values of $G$. This restriction should not be surprising because the original density function, $\rho$, is strictly periodic and in the limit of point masses represents a series of delta functions. In fact, the series of real-space lattice points at $a, 2a, 3a, \ldots$ for this simple 1D problem is simply the Bravais lattice vector defined by $\vec{R} = na\hat{x}$.

Extending to multiple dimensions, the restrictive relation between $\vec{G}$ and $\vec{R}$ is:

$$\vec{G}_n \cdot \vec{R}_m = 2\pi \times \text{integer}. \tag{1.8}$$

The vector $\vec{G}$ thus becomes critically important in the description of lattices—*the reciprocal lattice vector*. In the interest of brevity and following the lead of Ziman, we will not be "concerned here with mathematically pathological functions, and may use naive Fourier theory quite freely" (Ziman, 1972). As such, we will simply state the relations between reciprocal lattice translation vectors $\vec{b}_i$ and the direct lattice translation vectors $\vec{a}_i$ in 3D:

$$\vec{G} = k_1 \vec{b}_1 + k_2 \vec{b}_2 + k_3 \vec{b}_3, \tag{1.9}$$

where

$$\vec{b}_i = 2\pi \frac{\vec{a}_j \times \vec{a}_k}{\vec{a}_1 \cdot (\vec{a}_2 \times \vec{a}_3)}, \tag{1.10}$$

and $k_i$ are integers, and the denominator in Eq. (1.10) is the unit cell volume.

Once the $\vec{b}_i$ vectors are known, the reciprocal space can be populated with discrete points. We will focus on 2D graphene here. Analysis of the primitive translational vectors in Eq. (1.4) in the context of Eq. (1.8) reveals that we must have $\vec{b}_1 \perp \vec{a}_2$ and $\vec{b}_2 \perp \vec{a}_1$ and that

$$\vec{b}_1 = C_1 \left[ \frac{\sqrt{3}}{2} \hat{x} + \frac{3}{2} \hat{y} \right]$$

$$\vec{b}_2 = C_2 \left[ \frac{\sqrt{3}}{2} \hat{x} - \frac{3}{2} \hat{y} \right]. \tag{1.11}$$

The constants $C_1$ and $C_2$ must be equal to preserve the generality of Eq. (1.8), and given the magnitude of the vectors $|\vec{a}_i| = \sqrt{3}\tilde{a}$, we find:

$$C_1 = C_2 = \frac{4\pi}{a3\sqrt{3}}$$

$$\rightarrow \vec{b}_1 = \frac{2\pi}{\tilde{a}} \left[ \frac{1}{3} \hat{x} + \frac{1}{\sqrt{3}} \hat{y} \right]$$

$$\rightarrow \vec{b}_2 = \frac{2\pi}{\tilde{a}} \left[ \frac{1}{3} \hat{x} - \frac{1}{\sqrt{3}} \hat{y} \right]. \tag{1.12}$$

The resulting lattices, both direct (a) and reciprocal (b), for graphene
are shown in Fig. 1.11, as well as the respective translational vectors and
the so-called 1st Brillouin zone, which is hexagonal in shape. The reciprocal
lattice's primitive cell (i.e., 1st Brillouin zone) is established by connecting
lattice points with lines, which then define the shaded region of 2D space
closest to a given lattice point.

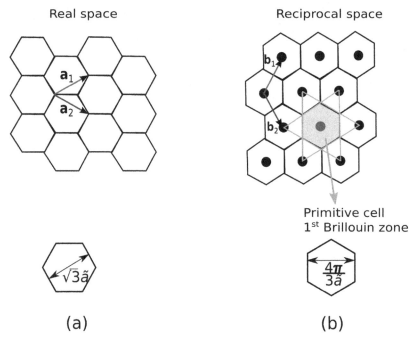

Fig. 1.11   (a) Direct graphene lattice. (b) Reciprocal graphene lattice. The primitive
cell of the reciprocal lattice is the 1st Brillouin zone. Translation vectors of both lattices
are also depicted.

Reciprocal space is often termed 'k-space', and we will use the terms
interchangeably. Reciprocal space is also useful in defining directions in a
crystal. For a given real-space lattice plane, the Miller indices $(k_1 k_2 k_3)$ are
vector coordinates (see Eq. (1.12)) of the shortest reciprocal lattice vector
normal to the plane. The Miller indices should not be confused with the
primary directions in the real-space lattice, which are denoted by square
brackets $[xyz]$.

## 1.4 Lattice Vibrations and Phonons

The description of lattice vibrations starts with the potential-energy/displacement relation of Eq. (1.2). When constructed as a linear chain of atoms, the individual potential energies from each compressed or expanded spring are summed to form the harmonic potential energy $U^{\mathrm{harm}}$:

$$U^{\mathrm{harm}} = \frac{1}{2}g \sum_n \{u[na] - u[(n+1)a]\}^2, \tag{1.13}$$

where the terms $na$ and $(n+1)a$ designate the spatial positions of the atoms. The force on an individual atom (at, say, location $na$) can be calculated from the spatial derivative of displacement at that location:

$$F = m\frac{d^2u(na)}{dt^2} = -\frac{\partial U^{\mathrm{harm}}}{\partial u(na)} = -g\{2u(na) - u[(n-1)a] - u[(n+1)a]\}, \tag{1.14}$$

where the factor 2 appearing in $2u(na)$ is the result of the fact that location '$na$' appears twice in the summation of Eq. (1.13) (once as $(n+1)a$ and then as $na$ as the sum proceeds). The '$na$' nomenclature becomes quite tedious in practice, and we therefore simplify the expression of Eq. (1.14) as:

$$m\frac{d^2u_n}{dt^2} = -g\{2u_n - u_{n-1} - u_{n+1}\}. \tag{1.15}$$

The solution of Eq. (1.15) requires boundary conditions, and the simplest are the so-called Born-von Karman type in which the ends of the 1D chain are attached as in a loop (see Fig. 1.12). We note that this 'loop' does not add a new dimension to the problem, as the number of atoms $N$ is assumed to be very large.

The Born-von Karman boundary conditions become:

$$\begin{aligned} u_N &= u_0 \\ u_{N+1} &= u_1. \end{aligned} \tag{1.16}$$

We assume a plane-wave solution for displacement at location $n$ as:

$$u_n(t) \sim \exp\{i(Kna - \omega t)\}, \tag{1.17}$$

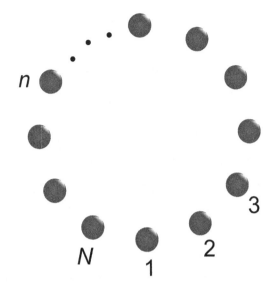

Fig. 1.12    1D chain of $N$ atoms with the Born-von Karman boundary condition.

where $K$ is the wavevector of the plane wave and is proportional to the inverse of wavelength. Application of Eq. (1.17) to the boundary conditions above yields:

$$u_{N+1} \sim \exp\left\{i\left[K\left(N+1\right)a - \omega t\right]\right\}$$

$$u_1 \sim \exp\left\{i\left[Ka - \omega t\right]\right\}$$

$$\rightarrow 1 = \exp\left[iKNa\right] \rightarrow KNa = 2\pi n, \tag{1.18}$$

where $n$ is an indexing integer. The final relation in Eq. (1.18) is of crucial importance, for it restricts the possible values of the wavevector $K$ that can 'fit' on the looped 1D chain. Of course, if the number of total atoms $N$ is large, then many wavevectors are possible. Defining the wavelength as $\lambda_n = aN/n$, the set of allowed wavevectors becomes:

$$K_n = \frac{2\pi n}{aN} = \frac{2\pi}{\lambda_n}. \tag{1.19}$$

Finally, we note that the minimum size of a wave (wavelength) is $\lambda_{\min} = 2a$, for any shorter waves would not have atoms to sustain them. Another way of explaining this characteristic is that any smaller wavelengths would have nodal positions (in the standing wave sense) that could be described by longer waves in which the nodal positions would exist on lattice sites, instead of between atoms. Consequently, the maximum unique wavevector is:

$$|K_{\mathrm{max,unique}}| = \frac{\pi}{a}. \tag{1.20}$$

This important restriction enables us to convert what is an infinite domain in real space (as $N \to \infty$) into a finite domain in reciprocal space ($K \in [-\pi/a, \pi/a]$), with the associated advantages of mathematical convenience. Importantly, this unique region of reciprocal space (or **k**-space) coincides with the 1st Brillouin zone.

We now return to the equation of motion, Eq. (1.15), and its solution. Substitution of the plane-wave function of Eq. (1.17) for position $na$ and incorporation of the discrete wavevectors $K_j$ produces:

$$
\begin{aligned}
- m\omega_j^2 e^{i(K_i na - \omega_j t)} &= -g \left[ 2 - e^{-iK_j a} - e^{iK_j a} \right] e^{i(K_j na - \omega_j t)} \\
&= -2g \left( 1 - \cos K_j a \right) e^{i(K_j na - \omega_j t)}. \tag{1.21}
\end{aligned}
$$

The resulting relationship between frequency and wavevector defines the *dispersion relation* of the lattice:

$$\omega_j(K_j) = \sqrt{\frac{2g(1 - \cos K_j a)}{m}} = 2\sqrt{\frac{g}{m}} \left| \sin(\tfrac{1}{2} K_j a) \right|. \tag{1.22}$$

The continuous form of this relation ($\omega(K)$, which we will use hereafter, dropping the subscript $j$) is sketched in Fig. 1.13. We note that the maximum frequency depends quite simply on the spring constant and atomic mass, as $\omega_{\mathrm{max}} = 2\sqrt{g/m}$.

The dispersion relation contains information pertinent to a wide range of material characteristics, from elastic constants to the scattering rates of phonons. We will discuss many of these in context throughout the remainder of the text. For now, we highlight the phase and group velocities:

$$\textit{phase velocity: } c = \frac{\omega}{K}, \tag{1.23}$$

$$\textit{group velocity: } v_g = \frac{\partial \omega}{\partial K}. \tag{1.24}$$

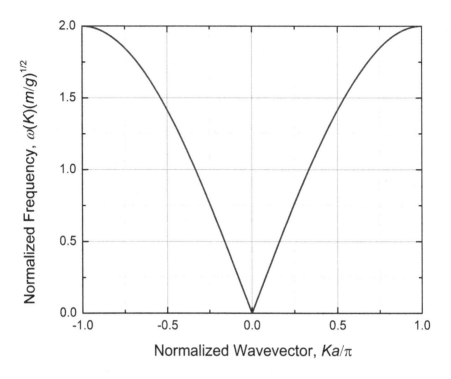

Fig. 1.13    Dispersion relation for a monatomic 1D chain of atoms.

Most of our interest will be given to the group velocity because it determines the rate of energy transport. Further, we will often focus on the long-wave limit $(K \to 0)$, for which:

$$\lim_{K \to 0} \omega = a\sqrt{\frac{g}{m}} \, |K|$$

$$\to \lim_{K \to 0} v_g = a\sqrt{\frac{g}{m}} = \left| \frac{\omega}{K} \right| = c. \tag{1.25}$$

In this limit, the group and phase velocities are equal, and they both are the same as the speed of sound in the solid. Therefore, the types of phonons that exhibit this behavior (other types are considered later) are termed *acoustic phonons*.

Thus far we have used strictly classical descriptions of mechanical vibrations to derive the vibrational spectrum of the lattice. However, to treat collections of vibrations (because a lattice can support many vibrational modes at the same time), we must transition to a quantum description. Nevertheless, we can retain the results from the classical harmonic oscillator solution above to define each *normal mode* in terms of a wavevector $K$ and frequency $\omega$ (i.e., the dispersion relation remains valid). A solution of the time-independent Schrödinger equation of quantum mechanics (see Eq. (1.31) in the next section) reveals that each mode can contain a set of energies described by:

$$\varepsilon_K = \left(N_K + \tfrac{1}{2}\right)\hbar\omega_K, \tag{1.26}$$

where $N_K$ represents the number of phonons with wavevector $K$, and the terminology $\omega_K$ is intended to signify the inherent relationship between frequency and wavevector embodied by the dispersion relation (Eq. (1.22)). The $\frac{1}{2}$ term in Eq. (1.26) accounts for the so-called zero-point energy whose derivation is available elsewhere (Ashcroft and Mermin, 1976, Appendix L). The term $N_K$ defines the average number of such excited modes of wavevector $K$, or the number of *phonons*, and is defined by Bose-Einstein statistics:

$$N_K = \frac{1}{\exp\left(\frac{\hbar\omega_K}{k_B T}\right) - 1}, \tag{1.27}$$

where $k_B$ is Boltzmann's constant, and $T$ is temperature. We will later use the symbol $f_{BE}^o$ as a synonym for $N_K$ (in attempt to maintain some consistency while also identifying various symbols that are used in the literature for the occupation number).

The connection between the quantum energy of Eq. (1.26) and the classical vibration amplitude is often elusive to new learners and is therefore included here to connect with mechanical intuition. Classically, each vibrational mode contains a combination of potential and kinetic energy that can be shown to be, on average, equal in magnitude by the virial theorem (Ashcroft and Mermin, 1976) such that:

$$\bar{\varepsilon}_{\text{classical}} = \sum_{\text{lattice}} m\,|\dot{u}|^2, \tag{1.28}$$

where the " $\dot{}$ " denotes time differentiation. For a simple lattice of $N$ atoms with one atom of mass $m$ per unit cell, the summation can be transformed to reciprocal space as:

$$\bar{\varepsilon}_{\text{classical}} = \sum_K Nm\omega_K^2 \left| \tilde{u}_K \right|^2, \qquad (1.29)$$

where $\left| \tilde{u}_K \right|$ is the amplitude of atomic displacement for a mode with wavevector $K$. Equating the summed term in Eq. (1.29) with the quantum version (Eq. (1.26)), the relationship between displacement amplitude and (quantized) energy becomes:

$$\left| \tilde{u}_K \right|^2 = \frac{\varepsilon_K}{Nm\omega_K^2}$$

$$= \frac{\left( N_K + \frac{1}{2} \right) \hbar}{Nm\omega_K}. \qquad (1.30)$$

This result should be intuitive, for it indicates that displacement amplitude increases with increasing occupation number and decreases with increasing frequency, both in the square-root sense. An illustration of phonon quantization, showing the relationship between allowed energies and atomic displacements, is shown in Fig. 1.14. For further details, the reader is referred to Ziman (1972).

Still remaining in our development is the extension of the foregoing principles of dispersion and energy to multiple dimensions and orientations of oscillations relative to the propagation direction (i.e., polarization). We defer these subjects to later chapters, when they can be developed in better context.

## 1.5   Free Electrons

Electronic behavior varies widely among different types of materials, from 'free' conduction in metals to virtually none in insulators. In this chapter we will consider only metals, and even then we will use the simplest approximation–free electron theory. Later chapters elucidate more complicated electronic structure.

The fundamental equation governing quantum particles is Schrödinger's equation, whose time-independent form is:

$$\frac{-\hbar^2}{2m_e} \nabla^2 \Psi(\vec{r}) + \mathcal{V}(\vec{r}) \Psi = E \Psi(\vec{r}), \qquad (1.31)$$

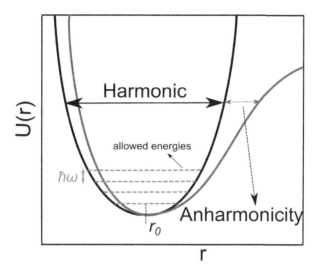

Fig. 1.14   Discrete energy levels depict phonon quantization. Successive energy levels are separated by $\hbar\omega$.

where $\Psi$ is the electron wavefunction, and $\mathcal{V}(\vec{r})$ represents a potential energy function that commonly represents the periodic ion field in a crystal. However, $\mathcal{V}(\vec{r}) = 0$ is assumed to make electrons 'free' in the present simplification. The wavefunction determines the probability per unit volume $P$ of finding an electron at position $\vec{r}$ according to:

$$P = \left|\Psi(\vec{r})^2\right| = \Psi(\vec{r})\Psi^*(\vec{r}), \qquad (1.32)$$

where the "$*$" denotes complex conjugation. Once again, we assume a plane-wave solution (in this case, a steady-state form):

$$\Psi_k(\vec{r}) = \frac{1}{\sqrt{V}}e^{i\vec{k}\cdot\vec{r}}, \qquad (1.33)$$

where $V$ is volume and $\vec{k}$ is the electron's wavevector.[1] Substitution into the governing equation yields an expression for the energy eigenvalue $E_k$:

$$E_k = \frac{\hbar^2 k^2}{2m_e}, \qquad (1.34)$$

---

[1]We will use the lowercase symbol $k$ for electrons, and the uppercase $K$ for phonons to signify the carrier type. The term **k**-space is generic and applies to either.

where $k = \left|\vec{k}\right|$. Equation (1.34) relates electron energy and wavevector and is the dispersion relation for electrons, analogous to Eq. (1.22) for phonons. In this case, the functional relationship is parabolic, $E_k \sim k^2$. Such parabolic dispersion relations (or bands) are common in real materials, even for those with complicated electronic structures.

The parabolic '$E - k$' relation suggests a connection between wavevector and momentum. The usual Newtonian expressions for momentum $p$ and energy become:

$$|p| = m_e |v| \; ; \; E = \frac{m_e v^2}{2} \rightarrow v = \sqrt{\frac{2E}{m_e}}$$

$$\rightarrow |p| = m_e \sqrt{\frac{2E}{m_e}} = \sqrt{2E m_e} = \sqrt{\hbar^2 k^2} = \hbar k \qquad (1.35)$$

$$\rightarrow \vec{p} = \hbar \vec{k}.$$

The final result indicates that the wavevector can be considered a surrogate for momentum.

The momentum of electrons is restricted to certain allowed states, as it was for phonons. For the free electron gas, we can determine these values by considering an electron in a cube (the so-called 'electron in a box' problem). The wavefunction and its corresponding probability functional in Eq. (1.32) are assumed to be spatially periodic (see Fig. 1.15), such that:

$$\Psi(x + L) = \Psi(x); \; \Psi(y + L) = \Psi(y); \; \Psi(z + L) = \Psi(z). \qquad (1.36)$$

Combining these periodic conditions with the plane-wave solution of Eq. 1.33 produces a set of allowable wavevectors:

$$e^{ik_x L} = e^{ik_y L} = e^{ik_z L} = 1$$

$$\rightarrow k_x = \frac{2\pi n_x}{L}, k_y = \frac{2\pi n_y}{L}, k_z = \frac{2\pi n_z}{L}, n_i = 1, 2, 3, \ldots \qquad (1.37)$$

This result should be familiar, for it is the same as that for phonons in the linear chain (Eq. (1.19)) for $L = aN$, the chain length. Therefore, allowable wavevectors are separated by $2\pi/L$ in reciprocal space; this characteristic will be useful in the next chapter in deriving the so-called density of states.

An important difference exists, however, between the manner in which the allowed wavevectors are populated for electrons and phonons. The latter can populate a state with a limitless number whose average (which need

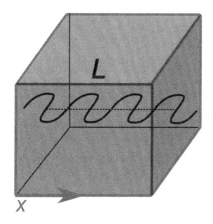

Fig. 1.15    Electron in a cube with a spatially periodic wavefunction.

*not* be an integer) is given by Eq. (1.27). Conversely, the electron occupation number of a given state is limited by the Pauli exclusion principle to be either 0 (not occupied) or 1 (occupied). Therefore, free electrons readily fill the reciprocal space until the number of carriers is exhausted.

Consider a material that contains $N'$ free electrons in a volume of real space $V$. The ratio of these is the electron density $\eta_e = N'/V$. Because each allowed state occupies a reciprocal-space volume of $(2\pi/L)^3$, the number of electrons can be expressed in terms of a spherical 'volume' of **k**-space as:

$$N' = 2\frac{\left(4\pi k^3/3\right)}{\left(2\pi/L\right)^3} = \frac{k_F^3}{3\pi^2}V, \tag{1.38}$$

where $k_F$ is called the Fermi wavevector and represents the largest occupied state at absolute zero temperature (the next chapter considers non-zero temperatures). The factor 2 in Eq. (1.38) accounts for the two electron spin states–up and down.

Other Fermi quantities can be easily derived from the Fermi wavevector:

$$\text{Fermi wavevector: } k_F = \left(3\pi^2\eta_e\right)^{1/3}, \tag{1.39}$$

$$\text{Fermi energy: } E_F = \frac{\hbar^2 k_F^2}{2m_e} = \frac{\hbar^2}{2m_e}\left(3\pi^2\eta_e\right)^{2/3}, \tag{1.40}$$

$$\text{Fermi velocity: } v_F = \frac{\hbar k_F}{m_e} = \frac{\hbar}{m_e}\left(3\pi^2\eta_e\right)^{1/3}, \qquad (1.41)$$

$$\text{Fermi temperature: } \theta_F = \frac{E_F}{k_B} = \frac{\hbar^2}{2m_e k_B}\left(3\pi^2\eta_e\right)^{2/3}. \qquad (1.42)$$

The Fermi energy $E_F$ is the most commonly used, and as shown in Eq. (1.40), can be calculated from the electron density. The Fermi velocity $v_F$ is also an important quantity because even though the electron velocities cover a very broad range, only states near the Fermi level are active in conduction because of the nearby availability of unoccupied states necessary to produce transport.

A sketch of the filled and empty energy levels is shown in Fig. 1.16. By convention, the zero energy datum is chosen to sit at the bottom of the conduction band, with non-conducting core electron states beneath. The electrons fill energies upward until they reach the Fermi energy and are contained in the solid by an energy barrier called the work function $\phi$, which is the difference between the vacuum energy $E_{\text{vac}}$ and Fermi energy $E_F$.

## 1.6   Example: 1D Atomic Chain with a Diatomic Basis

We choose a diatomic 1D chain of atoms as shown schematically in Fig. 1.17 to demonstrate a slightly more complicated situation than the monatomic chain of Section 1.4. The 2-atom basis produces an entirely separate phonon branch, as derived below.

For details of phonon analysis for linear chains, the reader is referred to Chapter 2 of Ziman (1972). We note that the definition of $a$ here, which is the distance between unit cells, is a bit different from Ziman's, which does not span a full unit cell but rather the distance between atoms within a cell. Here we include the essential elements starting again with the Lagrangian mechanics relation, $\mathbf{F} = m\ddot{\mathbf{u}} = -\nabla U^{\text{harm}}$ (cf., Eq. (1.14)), where $\mathbf{F}$ is the force on a particle of mass $m$ with displacement $\mathbf{u}$, and again $U^{\text{harm}}$ is the potential energy of the entire many-body system. Given the one-dimensional nature of the present formulation, we drop the spatial vector notation.

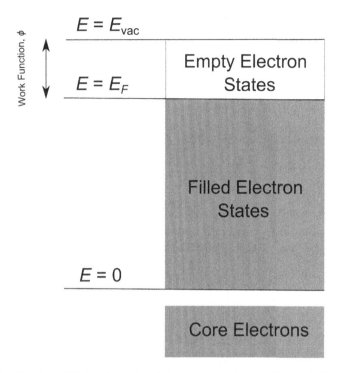

Fig. 1.16 Sketch of filled and empty electron energy states. The work function, $\phi$, is defined as the difference between the Fermi energy, $E_F$, and the vacuum energy, $E_{\text{vac}}$.

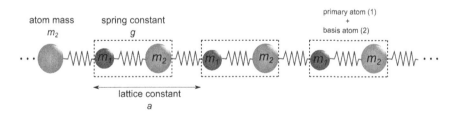

Fig. 1.17 Schematic of a 1D atomic chain with a two-atom basis.

In comparison to the monatomic chain, the index accounting for different atoms is more difficult when a basis atom is added (as well as any additional displacement dimensions not considered here); consequently, a matrix-based approach is required. To account for the discreteness of the system, we represent each atom's displacement as $u_{n,\alpha}$ where $\alpha$ is the basis index (1 or 2) and $n$ is the unit cell index. The equation of motion becomes (Ziman, 1972):

$$m_\alpha \ddot{u}_{n,\alpha} = -\sum_{m,\beta} \frac{\partial^2 U^{\text{harm}}}{\partial u_{n,\alpha} \partial u_{m,\beta}} u_{m,\beta} = -\sum_{m,\beta} \Phi_{n,\alpha}^{m,\beta} u_{m,\beta}, \qquad (1.43)$$

where $U^{\text{harm}}$ for this 2-atom basis is:

$$U^{\text{harm}} = \frac{1}{2}g \sum_n \left(u_{n,1} - u_{n,2}\right)^2 + \left(u_{n,2} - u_{n+1,1}\right)^2. \qquad (1.44)$$

The matrix $\Phi_{n,\alpha}^{m,\beta}$ (hereafter called the 'force constants matrix') contains the interatomic force constants between each atom pair (i.e., $(n,\alpha)$ and $(m,\beta)$). Inspection of Eqs. (1.43) and (1.44) reveals:

$$\Phi_{n,1}^{n,1} = \Phi_{n,2}^{n,2} = 2g$$

$$\Phi_{n,2}^{n,1} = \Phi_{n,1}^{n,2} = \Phi_{n,1}^{n-1,2} = \Phi_{n,2}^{n+1,1} = -g. \qquad (1.45)$$

Recognizing the symmetry of the problem (i.e., that all unit cells are identical) and using a left-to-right numbering scheme, the force constants matrix becomes:

$$\Phi = \begin{bmatrix} 2g & -g \\ -g & 2g \end{bmatrix}. \qquad (1.46)$$

From the translational symmetry of the chain, the unit cell indices $n$ and $m$ can be replaced by 0 and $p$, respectively, where $p$ is simply an index that begins at 0 and increases in unit steps away from the cell of interest (i.e., 0). The Fourier transform of this matrix becomes the so-called dynamical matrix of lattice dynamics analysis (Young and Maris, 1989):

$$D_\alpha^\beta = \frac{1}{\sqrt{m_\alpha m_\beta}} \Phi_{0,\alpha}^{p,\beta} e^{i\vec{K}\cdot\vec{r}_p}$$

$$= \begin{pmatrix} \frac{2g}{m_1} & \frac{-g}{\sqrt{m_1 m_2}}\left(1 + e^{-iKa}\right) \\ \frac{-g}{\sqrt{m_1 m_2}}\left(1 + e^{+iKa}\right) & \frac{2g}{m_2} \end{pmatrix}, \qquad (1.47)$$

where $\vec{r}_p$ is the distance between the unit cells of the pair of atoms under consideration and implied summation applies to the index $p$. The dynamical matrix emerges as part of the governing equation of motion (Eq. (1.43)) cast in frequency space:

$$\omega^2 \tilde{u}_\alpha(K) = \frac{1}{\sqrt{m_\alpha m_\beta}} \Phi^{p\beta j}_{0\alpha i} e^{iK \cdot \vec{r}_p} \tilde{u}_\beta(K) = D^\beta_\alpha \tilde{u}_\beta(K), \qquad (1.48)$$

where $\tilde{u}$ is the amplitude of displacement. The so-called secular equation emerges from the foregoing expression and is used the extract the eigenvalues $\omega^2$:

$$\det \left| \mathbf{D} - \omega^2 \mathbf{I} \right| = 0, \qquad (1.49)$$

where $\mathbf{I}$ is the identity matrix (2 by 2 in this case). The solution of Eq. (1.49) provides a form of the dispersion relation:

$$\omega^4 - 2g \left( \frac{1}{m_1} + \frac{1}{m_2} \right) \omega^2 + \frac{4g^2}{m_1 m_2} \sin^2 \left( \frac{Ka}{2} \right) = 0. \qquad (1.50)$$

Using quadratic reduction, the foregoing result can be solved for $\omega^2$:

$$\omega(K)^2 = g \left( \frac{1}{m_1} + \frac{1}{m_2} \right) \pm g \sqrt{\left( \frac{1}{m_1} + \frac{1}{m_2} \right)^2 - \frac{4}{m_1 m_2} \sin^2 \left( \frac{Ka}{2} \right)}. \qquad (1.51)$$

The '$\pm$' term in Eq. (1.51) produces the peculiarity of having two possible branches. The lower branch (defined as having the lower frequency, represented by $\omega_-$) is the acoustic branch and is equivalent to that derived for the monatomic chain (Eq. (1.22)). The upper branch ($\omega_+$), called the *optical branch*, is new and represents generally out-of-phase vibrations between neighboring atoms (i.e., the displacements of neighboring atoms are nearly equal and opposite). The limiting forms at the Brillouin zone origin and edges for both branches are:

$$\lim_{K \to 0} \omega_-(K) = Ka \sqrt{\frac{g\mu}{2m_1 m_2}} \; ; \quad \lim_{K \to 0} \omega_+(K) = \sqrt{\frac{2g}{\mu}}, \qquad (1.52)$$

$$\omega_- \left( K = \frac{\pi}{a} \right) = \sqrt{\frac{2g}{m_2}} \; ; \quad \omega_+ \left( K = \frac{\pi}{a} \right) = \sqrt{\frac{2g}{m_1}}, \qquad (1.53)$$

where $\mu = (1/m_1 + 1/m_2)^{-1}$, and $m_2$ is the heavier of the two masses.

The two branches of the dispersion are shown in Fig. 1.18. The curves reveal an energy band gap between the branches that grows with increasing contrast between the two atomic masses. Moreover, the shape of the optical branch takes a generally flat character, suggesting that the group velocity $(d\omega/dK)$ is relatively small. Consequently, optical phonons are often neglected in the calculation of thermal conductivity, in favor of the acoustic branch.

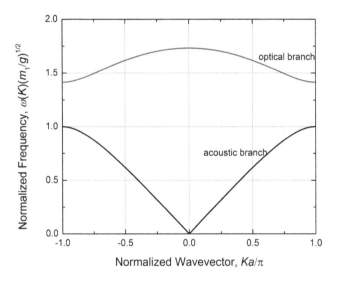

Fig. 1.18   Normalized frequency as a function of normalized wavevector for a diatomic 1D chain with $m_2 = 2m_1$.

Finally, we address a common source of confusion related to the application of the diatomic dispersion results to the case of $m_1 = m_2$ (i.e., the monatomic case). Figure 1.19 shows the dispersion curve for the range $K \in \{-\pi/a, \pi/a\}$. Notably, the solution still predicts the presence of an optical mode, and thus one might wonder whether the optical mode is simply an artifact of mathematics. However, careful inspection reveals that the condition $m_1 = m_2$ produces one-eighth of a full sine wave for the acoustic branch in the range $K \in \{0, \pi/a\}$ instead of the usual quarter sine wave (cf., Figs. 1.13 and 1.18). The reason for this change is that the diatomic analysis uses a lattice constant $a = 2\tilde{a}$ that is exactly twice as large as that for the monatomic case ($a = \tilde{a}$) for this special case of $m_1 = m_2$. Therefore, the range of $K$ sampled in Fig. 1.19 is only half that of Fig. 1.13. The missing portion of the dispersion curve is actually contained in the optical branch, as shown in Fig. 1.19, which spans $K \in \{0, 2\pi/a\}$ and shows the completion of the quarter sine wave by the 'virtual' optical branch. This result provides an example of the importance of defining the primitive unit cell, from which the 1st Brillouin zone derives, as the *smallest* symmetric region necessary to fill exactly all space through the translation vector $\vec{R}$.

## 1.7 Conclusion

This chapter has laid a foundation in crystallography and the fundamentals of phonons and electrons, albeit in a highly idealized and simplified form. Often the mathematics of these fundamentals can obscure more intuitive or at least more familiar understanding. For example:

- The speed of sound in silicon (Si) is approximately 6400 m/s. With its nearest neighbor distance of 0.235 nm and atomic mass of 28.0855 g/mol, it is a straightforward exercise to estimate the spring constant from the $K \to 0$ limit (Eq. (1.25)) as $g = 35$ N/m, which is remarkably similar to the actual value (Zhang *et al.*, 2007).
- Many metals have Fermi energies near $E_F = 5$ eV. The corresponding Fermi velocity is $v_F = \sqrt{2E_F/m_e} \approx 10^6$ m/s, which is roughly two orders of magnitude less than the speed of light, $c_0 = 2.99792458 \times 10^8$ m/s.

Lastly, we include here a brief glossary of some the concepts covered in this chapter:

- **Primitive Cell:** A region of space that is closer to one point than any others.
- **Bravais Lattice:** A distribution of points in space that defines a repeating pattern.
- **1st Brillouin Zone:** The primitive cell of the reciprocal lattice.
- **Miller Indices:** Coordinates ($hkl$) of the shortest reciprocal lattice vector normal to a given real-space plane.
- **Group Velocity:** The speed at which phonons carry energy in a lattice (see Eq. (1.24)).

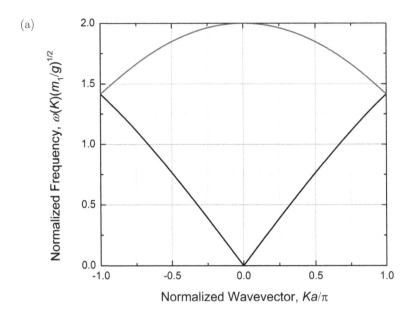

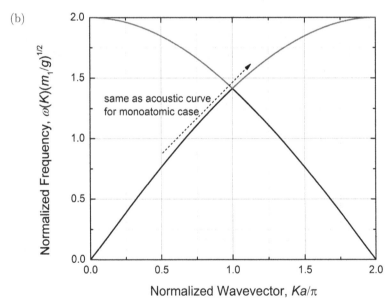

Fig. 1.19  Normalized frequency as a function of normalized wavevector for a diatomic 1D chain, for the special case of $m_1 = m_2$. (a) The usual range $K \in \{-\pi/a, \pi/a\}$. (b) The range $K \in \{0, 2\pi/a\}$.

- **Normal Mode:** A lattice wave that is characterized by a branch, wavevector, and frequency (and, later, polarization).
- **Phonon:** a quantized lattice vibration (i.e., one that can take on only a discrete energy, $\hbar\omega$).
- **Acoustic Phonons:** Phonons that determine the speed of sound in a solid and are characterized by $\omega \sim K$ as $K \to 0$.
- **Optical Phonons:** Phonons that have flat dispersion, low group velocity, and are characterized by non-zero $\omega$ as $K \to 0$.
- **Occupation Number:** The number of carriers with a given wavevector.

**Example Problems**

---

*Problem 1.1: Graphene reciprocal lattice*

The primitive lattice vectors $(\vec{a}_1, \vec{a}_2)$ of graphene are given by:

$$\vec{a}_1 = \frac{3}{2}\tilde{a}\hat{x} + \frac{\sqrt{3}}{2}\tilde{a}\hat{y},$$

$$\vec{a}_2 = \frac{3}{2}\tilde{a}\hat{x} - \frac{\sqrt{3}}{2}\tilde{a}\hat{y},$$

where $\tilde{a}$ is the C-C bond length. Calculate the reciprocal lattice vectors $\vec{b}_1$, $\vec{b}_2$ of graphene. Show that the primitive unit cell of the reciprocal lattice (also known as the 1st Brillouin zone) is a hexagon with a side length of $\frac{4\pi}{3\sqrt{3}\tilde{a}}$.

*Solution*

The primitive lattice vectors are:

$$\vec{a}_1 = \frac{3}{2}\tilde{a}\hat{x} + \frac{\sqrt{3}}{2}\tilde{a}\hat{y}, \quad \vec{a}_2 = \frac{3}{2}\tilde{a}\hat{x} - \frac{\sqrt{3}}{2}\tilde{a}\hat{y}, \quad \vec{a}_3 = c\hat{z}, \qquad (1.54)$$

where $c$ is an arbitrarily large constant (no periodicity exists in the $z$ direction). The reciprocal lattice vectors are then given by:

$$\vec{b}_1 = 2\pi \frac{\vec{a}_2 \times \vec{a}_3}{\vec{a}_1 \cdot (\vec{a}_2 \times \vec{a}_3)}$$

$$= \frac{2\pi}{\tilde{a}}\left(\frac{1}{3}\hat{x} + \frac{1}{\sqrt{3}}\hat{y}\right), \qquad (1.55)$$

$$\vec{b}_2 = 2\pi \frac{\vec{a}_3 \times \vec{a}_1}{\vec{a}_1 \cdot (\vec{a}_2 \times \vec{a}_3)}$$

$$= \frac{2\pi}{\tilde{a}}\left(\frac{1}{3}\hat{x} - \frac{1}{\sqrt{3}}\hat{y}\right), \qquad (1.56)$$

$$\vec{b}_3 = 2\pi \frac{\vec{a}_1 \times \vec{a}_2}{\vec{a}_1 \cdot (\vec{a}_2 \times \vec{a}_3)}$$

$$= \frac{2\pi}{c}\hat{z}, \qquad (1.57)$$

where $\vec{b}_3 \to 0$ as $c \to \infty$ indicating that the reciprocal lattice is two-dimensional. Figure 1.20 shows the process involved in construction of the reciprocal lattice. The $\Gamma$ point of the reciprocal lattice is joined to its six nearest neighbors given by the points $\vec{b}_1$ (point A), $\vec{b}_2$ (point C), $\vec{b}_1 + \vec{b}_2$ (point B), $-\vec{b}_1$ (point D), $-\vec{b}_2$ (point F) and $-\vec{b}_1 - \vec{b}_2$ (point E). Perpendicular bisectors (red dotted lines in Fig. 1.20) are then drawn for each of these six line segments $\Gamma$A, $\Gamma$B, $\Gamma$C, $\Gamma$D, $\Gamma$E and $\Gamma$F. The region of intersection of these perpendicular bisectors forms the hexagonal Brillouin zone of graphene.

The side of the hexagon can be obtained from simple trigonometry. $\angle M\Gamma K = \frac{1}{2}\angle M\Gamma B = 30°$. Thus $MK = \Gamma M/\sqrt{3} = |\vec{b}_1|/2\sqrt{3} = 2\pi/3\sqrt{3}\tilde{a}$. One side of the hexagonal Brillouin zone is $2MK = 4\pi/3\sqrt{3}\tilde{a}$.

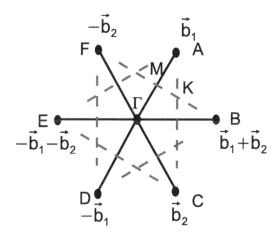

Fig. 1.20  Construction of the graphene Brillouin zone.

**Problem 1.2: Dispersion relation for a 1D chain**

(a) Consider a monoatomic 1D chain with nearest neighbor interactions. Assume a spring constant $g = 25$ N/m, atomic mass $m = 28$ amu and a lattice spacing of 5 Å. Calculate the sound velocity and the maximum possible phonon frequency.

(b) Now we generalize the monoatomic chain to include long-range interactions among atoms. Assume that the spring constant between two atoms separated by a distance of $ja$ is given by $g_j$ where $j$ is an index that can take values $1, 2, 3$ and so on. Show that the new dispersion relation is given by:

$$\omega = 2\sqrt{\sum_j \frac{g_j \sin^2(\frac{1}{2}jKa)}{m}}.$$

*Solution*

(a) Sound velocity is given by:

$$c = a\sqrt{\frac{g}{m}} = 11594 \text{ m/s}, \tag{1.58}$$

where mass is converted to kg ($1$ amu $= 1.6605 \times 10^{-27}$ kg). The maximum phonon frequency is given by:

$$\omega_{\max} = 2\sqrt{\frac{g}{m}} = 4.63 \times 10^{13} \text{ rad/s}, \tag{1.59}$$

(b) When interactions are considered between all pairs of atoms, the equation of motion for the atom at position $na$ is given by:

$$m\frac{d^2u(na)}{dt^2} = \sum_{j=1}^{j=\infty} g_j(u[(n+j)a] - u(na)) - g_j(u(na) - u[(n-j)a])$$

$$= \sum_{j=1}^{j=\infty} g_j(u[(n+j)a] - 2u(na) + u[(n-j)a]). \tag{1.60}$$

Substituting the plane wave solution $u(na) = \exp(i(Kna - \omega t))$ into the above equation produces:

$$-m\omega^2 = \sum_{j=1}^{j=\infty} g_j(\exp{(ijKa)} - 2 + \exp{(-ijKa)})$$

$$= \sum_{j=1}^{j=\infty} -4g_j \sin^2\left(\frac{jKa}{2}\right). \tag{1.61}$$

Thus the generalized dispersion relation is given by:

$$\omega = 2\sqrt{\sum_{j=1}^{j=\infty} \frac{g_j \sin^2(\frac{1}{2}jKa)}{m}}. \tag{1.62}$$

## Problem 1.3: Kinetic energy of the free electron gas

Obtain an expression for the total kinetic energy of the free electron gas at $T = 0$ K. Express your answer in terms of the Fermi energy $E_F$ and the total number of electrons $N$.

*Solution*

The following expressions for Fermi wavevector and Fermi energy were derived in the chapter:

$$k_F = (3\pi^2\eta_e)^{1/3}, \quad E_F = \frac{\hbar^2 k_F^2}{2m_e} = \frac{\hbar^2}{2m_e}(3\pi^2\eta_e)^{2/3}. \tag{1.63}$$

The volume of a spherical shell in **k**-space with radius $k$ and thickness $dk$ is given by $4\pi k^2 dk$. The number of states in this shell is given by:

$$dN = 2\frac{4\pi k^2 dk}{(2\pi/L)^3}, \tag{1.64}$$

where the factor 2 accounts for spin degeneracy. The energy of each state on the spherical shell of radius $k$ is $\hbar^2 k^2/2m_e$. The total energy of electrons is obtained by integrating up to the maximum wavevector $k_F$.

$$E = \int\limits_0^{k_F} \left( \frac{\hbar^2 k^2}{2 m_e} \right) \left( 2 \frac{4\pi k^2 dk}{(2\pi/L)^3} \right)$$

$$= \frac{\hbar^2 L^3}{2 m_e \pi^2} \int\limits_0^{k_F} k^4 dk$$

$$= \frac{\hbar^2 L^3 k_F^5}{10 m_e \pi^2}$$

$$= \frac{3}{5} \underbrace{\left( \frac{k_F^3 L^3}{3\pi^2} \right)}_{N} \underbrace{\left( \frac{\hbar^2 k_F^2}{2 m_e} \right)}_{E_F}$$

$$= \frac{3}{5} N E_F. \tag{1.65}$$

---

**Problem 1.4: Phonon bandgap in a diatomic chain**

Consider the diatomic chain (discussed in Section 1.6) with atomic masses $m_1$ and $m_2$. Assume that the spring constant is $g$ for all the bonds. At what point in the Brillouin zone is the bandgap (difference between the optical and acoustic branch frequencies) a minimum? Obtain an expression for the non-dimensional bandgap (normalized by $\sqrt{g/m_1}$) as a function of the mass ratio $m_2/m_1$. Use the online Chapter 1 CDF tool[2] to observe the changes in shape of the acoustic and optical branches for varying mass ratio.

*Solution*

Observation of the acoustic and optical branches of a diatomic chain reveals that the bandgap is minimum at the edge of the Brillouin zone. The dispersion relation for a 1D diatomic chain of atoms is given by (see Eq. (1.51)):

$$\omega(K)^2 = g \left( \frac{1}{m_1} + \frac{1}{m_2} \right) \pm g \sqrt{ \left( \frac{1}{m_1} + \frac{1}{m_2} \right)^2 - \frac{4}{m_1 m_2} \sin^2 \left( \frac{Ka}{2} \right) }.$$

$$\tag{1.66}$$

---

[2]See http://nanohub.org/groups/cdf_tools_thermal_energy_course/wiki

The angular frequency at the edge of the Brillouin zone is obtained by substituting $K = \pi/a$:

$$\omega_- \left(K = \frac{\pi}{a}\right) = \sqrt{\frac{2g}{m_2}}, \quad \omega_+ \left(K = \frac{\pi}{a}\right) = \sqrt{\frac{2g}{m_1}}. \tag{1.67}$$

The non-dimensional bandgap is given by:

$$\frac{\omega_+ \left(K = \frac{\pi}{a}\right) - \omega_- \left(K = \frac{\pi}{a}\right)}{\sqrt{\frac{g}{m_1}}} = \sqrt{2} \left(1 - \sqrt{\frac{m_1}{m_2}}\right), \quad (m_2 > m_1). \tag{1.68}$$

From the above expression, the bandgap increases with increasing mismatch between the masses $m_1$ and $m_2$. Figure 1.21 shows snapshots of the dispersion curves from the online Chapter 1 CDF tool. The optical branch flattens and the maximum frequency of the acoustic branch reduces for increasing $m_2/m_1$.

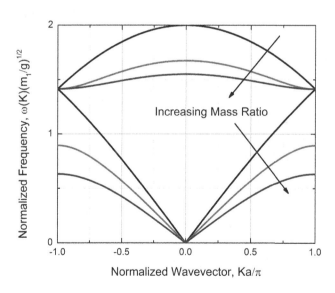

Fig. 1.21   Dispersion curves of the diatomic chain for increasing $m_2/m_1$.

Chapter 2

# Carrier Statistics

## 2.1 Introduction

Before proceeding further into details of nanoscale structure and energy transport, we first consider in this chapter an important distinction between nano*science* and nano*technology*. The former has been practiced for a century, ever since the nature of atomic structure was uncovered by the likes of Ernest Rutherford, Niels Böhr, and Marie Curie. The foundations of nanotechnology were similarly laid by researchers in the physical sciences, but nanotechnology is almost always characterized by a unique, *collective* behavior of an ensemble of nanoscale objects. In other words, nanotechnology encompasses phenomena that occur because of unique subcontinuum effects and that also can be directed towards a useful technological purpose.

The discipline of statistical mechanics provides the tools to achieve descriptions of large assemblies of nanoscale objects and is the primary subject of this chapter. Once again, we provide here only the basic essentials, while directing the motivated reader to more comprehensive coverage in topical books by, for example, Chandler (1987) and Laurendeau (2005).

## 2.2 Statistical Ensembles

A collection of energetic particles can be characterized by its number of particles $N$, volume $V$, and energy $E$. The collection can exist in a variety of states in which the foregoing variables may change upon application of a suitable perturbation. To analyze the diversity of states, we invoke the fundamental premise of statistical mechanics (Chandler, 1987):

> During a measurement (e.g., of temperature), every possible state does in fact occur such that observed properties are averages of all possible states.

The statistics of any ensemble can be described by defining its allowed states, and each state can be represented by $\nu = (N, V, E)$. We define $\Omega_\nu$ as the number of possible arrangements of the ensemble that can produce the state $\nu$ within $dE$ of energy $E$. The fundamental *statistical assumption* that all states are equally probable then implies the probability that a particle will be in a state $\nu$ is:

$$P_\nu = \frac{1}{\Omega_\nu}. \tag{2.1}$$

The number of states also provides insight into the randomness that is possible in a given ensemble. Such disorder forms the essence of the thermodynamic property called entropy, which was derived by Boltzmann as:

$$S = k_B \ln(\Omega_\nu). \tag{2.2}$$

This result, when combined with the Maxwell relations (Laurendeau, 2005), produces a statistical definition for temperature:

$$\frac{1}{T} = \left(\frac{\partial S}{\partial E}\right)_{N,V}, \tag{2.3}$$

or

$$\beta = \frac{1}{k_B T} = \left(\frac{\partial \ln \Omega_\nu}{\partial E}\right)_{N,V}, \tag{2.4}$$

where the term $\beta$ is a common thermodynamic expression for the inverse of 'thermal energy' $k_B T$.

Various permutations and restrictions can be applied to an ensemble in order to change its state. Energy and particle number are the most important such properties for our purposes. If these properties are allowed to vary within an ensemble, then the probability of a state $\nu$ can be shown to be Laurendeau (2005):

$$P_\nu = \frac{\exp(-\beta E_\nu + \beta \mu N_\nu)}{\Xi}, \tag{2.5}$$

where

$$\Xi = \sum_\nu \exp(-\beta E_\nu + \beta \mu N_\nu), \tag{2.6}$$

and where $\mu$ is the electrochemical potential; the latter equation defines the ensemble's partition function $\Xi$, which is essentially a normalizing factor to ensure that the total probability sums to unity. Further, the quantification of ensemble statistics allows the calculation of averages through by taking 'moments' of probability. For example, an average energy can be calculated as:

$$\langle E \rangle = \sum_\nu P_\nu E_\nu. \tag{2.7}$$

An ensemble in which both the energy and number of particles are allowed to vary is termed the *grand canonical ensemble*. A given state will have $N_1$ particles each with energy $E_1$, $N_2$ particles with energy $E_2$, and so on. Their average total energy and particle numbers can be expressed using Eqs. (2.5)–(2.7) as:

$$\langle E \rangle = -\left( \frac{\partial \ln \Xi}{\partial \beta} \right)_{\beta\mu, V}, \tag{2.8}$$

$$\langle N \rangle = \left( \frac{\partial \ln \Xi}{\partial (\beta\mu)} \right)_{\beta, V}. \tag{2.9}$$

The statistics of the particles of interest here can be described in this manner, noting that:

- Bose-Einstein statistics, which govern phonons and photons, allow integer occupation numbers $N_\nu = 0, 1, 2, 3, \ldots$.
- Fermi-Dirac statistics, which govern electrons, allow only binary occupation numbers $N_\nu = 0$ or 1, as a result of the Pauli exclusion principle (Ashcroft and Mermin, 1976).
- Both of the above statistics converge at very high particle energies (relative to the thermal energy $k_B T$) to classical Maxwell-Boltzmann occupation statistics.

The resulting *average occupation numbers* $\langle N_\nu \rangle$, despite these differences, can be expressed in a general form as:

$$\boxed{\begin{aligned} f_i^o &= \frac{1}{e^{\frac{E_i - \mu}{k_B T}} + \gamma} \\ \gamma &= 1 \text{ (Fermi-Dirac, } i = FD) \\ \gamma &= -1 \text{ (Bose-Einstein, } i = BE) \\ \gamma &= 0 \text{ (Maxwell-Boltzmann, } i = MB) \end{aligned}}, \tag{2.10}$$

where the equilibrium 'distribution function' $f_i^o$ is synonymous with the average occupation number and is used hereafter. The Maxwell-Boltzmann distribution ($\gamma = 0$) represents the limit ($E_i - \mu$) $\gg k_B T$ for both Fermi-Dirac and Bose-Einstein distributions.

The electrochemical potential depends on the enumeration of carriers. It is zero for systems with an indefinite number of carriers (i.e., phonons and photons) (Pathria and Beale, 2011). The Helmholtz free energy can be expressed as $A = \mu N - pV$, and if $N$ is unbounded, the equilibrium number of particles $N$ must be determined by minimizing $A$ with respect to $N$, which by definition requires ($\partial A/\partial N)_V = \mu = 0$. For phonons and thermal photons (which are both called 'bosons' because they follow Bose-Einstein statistics) $\mu$ is zero because these particles can be created or destroyed at random without a change in the electrochemical potential. For electrons in metals, which have a finite number of carriers, $\mu$ can be approximated in terms of the Fermi energy $E_F$ and temperature as (Ashcroft and Mermin, 1976):

$$\mu \approx E_F \left[ 1 - \frac{\pi^2}{12} \left( \frac{k_B T}{E_F} \right)^2 \right], \tag{2.11}$$

where the Fermi energy represents the highest occupied energy at absolute zero temperature and was expressed for a free-electron gas previously in Eq. (1.40) (See also Eq. (5.20) of Zhang (2007)).

As an example, consider a distribution of phonons from which we choose those with a particular frequency $\omega_\nu$. While the energy of each of these phonons has already been shown to be $\hbar\omega_\nu$, the actual number of such phonons at a given temperature must be determined from statistics. In accord with the energy of a given phonon mode derived in the previous chapter (see Eq. (1.26)), the average (or expected) energy of all phonons at this frequency is:

$$\langle E_\nu \rangle = \hbar\omega_\nu \left\{ f_{BE}^o(\omega_\nu, T) + \frac{1}{2} \right\}$$

$$= \hbar\omega_\nu \left\{ \left[ \exp\left( \frac{\hbar\omega_\nu}{k_B T} \right) - 1 \right]^{-1} + \frac{1}{2} \right\}, \tag{2.12}$$

where the 1/2 term accounts for zero-point energy (i.e., energy at zero absolute temperature). Figure 2.1 shows the resulting variation of average energy for phonons at frequencies of $\omega_\nu = 10^{13}$ and $\omega_\nu = 10^{14}$ rad/s. The figure also contains a comparison with the classical energy $k_B T$ that arises

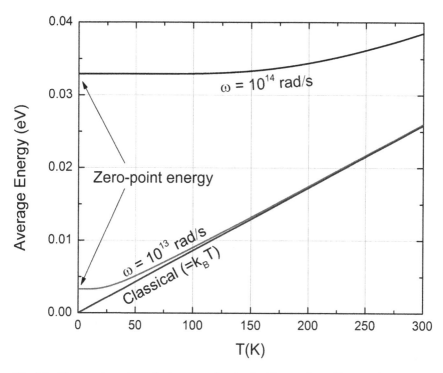

Fig. 2.1 Temperature-dependent energy of a classical harmonic oscillator and two quantum harmonic oscillators at $\omega = 10^{13}$ rad/s and $\omega = 10^{14}$ rad/s. The zero-point energy for quantum oscillators is on the $y$-axis.

from the equipartition theorem (Laurendeau, 2005). In this case, half of the 'thermal energy' $k_B T$ comes from the average kinetic energy in a single direction ($k_B T/2$), while the other half originates from the average potential energy in the bonds.

Clearly, temperature is intimately related to energy in both classical and quantum systems. For a simple (classical) harmonic oscillator, the relationship is direct: $\langle E_{\text{tot}} \rangle = k_B T$. For a quantum harmonic oscillator, the temperature dependence is contained within the occupation number: $\langle E_{\text{tot}} \rangle = \hbar\omega \left[ f^o_{BE}(\omega, T) + 1/2 \right]$. To complicate matters further, an atomic lattice can support many harmonic oscillators (according to the dispersion relation), and we need to sum (average) over all of their frequencies to find temperature. The results of Fig. 2.1 suggest that phonons oscillating at frequencies of $\omega = 10^{13}$ rad/s (and lower) can be approximated by classical statistics for all except cryogenic temperatures (i.e., below liquid nitrogen temperature, 77 K), whereas phonons at $\omega = 10^{14}$ rad/s require quantum statistics, even well above room temperature. The general guideline for assessing whether a classical approximation is appropriate is to evaluate $\chi = \frac{\hbar\omega}{k_B T}$. For example, at room temperature:

$$\frac{\hbar \times 10^{13} \text{ rad/s}}{k_B T} = 0.25 \ (\chi < 1, \text{ classical approximation is reasonable})$$

$$\frac{\hbar \times 10^{14} \text{ rad/s}}{k_B T} = 2.5 \ (\chi > 1, \text{ classical approximation is not reasonable}).$$

## 2.3 Phonon Density of States

The need to quantify the number of states around a certain energy or wavevector is common in the integral analysis of phonons. The associated quantity is called the *density of states* and describes the number of allowable phonon states per unit 'volume' (i.e., length in 1D, area in 2D, true volume in 3D) per unit energy or wavevector, depending on context. Recalling the restriction on allowable wavevectors from the looped 1D chain example (see Eq. (1.19)), we notice that allowable wavevectors are separated by a k-space increment of $2\pi/L$, where $L = aN$ ($N$ is the number of primitive unit cells; for cells with a single atom per unit cell $N$ is therefore the number atoms). We find a similar result for 2D and 3D lattices for which each allowable wavevector occupies a k-space 'volume' of $(2\pi/L)^d$, where $d$ is dimensionality.

The number of states is calculated by forming the ratio of a smooth (i.e., circular in 2D, spherical in 3D) k-space 'volume' to that of an individual state, as shown in Fig. 2.2. One subtlety of the foregoing development is that **K** can take both positive and negative values, and therefore with $K$ defined as the absolute magnitude of **K** (i.e., $K = |\mathbf{K}|$), one allowable wavevector exists for each increment of $\pi/L$ in the 1D k-space (which, by definition, is strictly positive) as a special case. Therefore, the number of allowed phonon states from 0 to $K$ (which, recall, is the absolute magnitude of **K**) for 1-, 2-, and 3-dimensional systems is:

$$N_{1D} = \frac{2K}{2\pi/L}, \tag{2.13}$$

$$N_{2D} = \frac{\pi K^2}{4\pi^2/L^2}, \tag{2.14}$$

$$N_{3D} = \frac{4\pi K^3/3}{8\pi^3/L^3}. \tag{2.15}$$

Once this number of allowed modes or phonons is known, the density of such states (per unit wavevector and real-space 'volume') can be expressed as:

$$D_{1D}(K) = \frac{1}{L}\frac{dN_{1D}}{dK} = \frac{1}{\pi}, \tag{2.16}$$

$$D_{2D}(K) = \frac{1}{L^2}\frac{dN_{2D}}{dK} = \frac{K}{2\pi}, \tag{2.17}$$

$$D_{3D}(K) = \frac{1}{L^3}\frac{dN_{3D}}{dK} = \frac{K^2}{2\pi^2}. \tag{2.18}$$

The density of states is often described with respect to phonon frequency instead of wavevector. This transformation is made quite readily using the chain rule and the definition of phonon group velocity ($v_g(\omega) = d\omega/dK$):

$$D_{1D}(\omega) = \frac{1}{L}\frac{dN_{1D}}{d\omega} = \frac{1}{L}\frac{dN_{1D}}{dK}\frac{dK}{d\omega} = \frac{1}{v_g(\omega)\pi}, \tag{2.19}$$

$$D_{2D}(\omega) = \frac{1}{L^2}\frac{dN_{2D}}{d\omega} = \frac{1}{L^2}\frac{dN_{2D}}{dK}\frac{dK}{d\omega} = \frac{K(\omega)}{2\pi v_g(\omega)}, \tag{2.20}$$

$$D_{3D}(\omega) = \frac{1}{L^3}\frac{dN_{3D}}{d\omega} = \frac{1}{L^3}\frac{dN_{3D}}{dK}\frac{dK}{d\omega} = \frac{K(\omega)^2}{2\pi^2 v_g(\omega)}, \tag{2.21}$$

where, as we have shown previously (see Eq. (1.22)), the phonon group velocity generally depends on frequency (and thus the wavevector through the dispersion relation).

At this point in the development, the $K(\omega)$ relations above are often simplified for phonon transport through the Debye approximation, which assumes linear dispersion $\omega = v_{g,\text{ave}}K$ and places an upper bound $\omega_D$ on frequency in order to match the total number of possible phonon states.

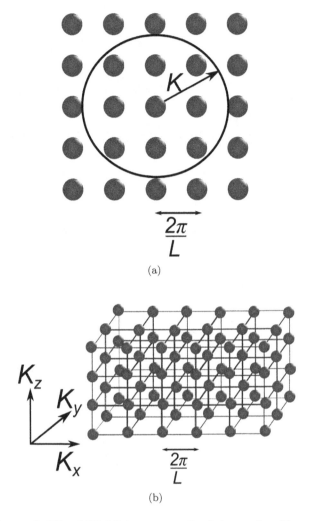

(a)

(b)

Fig. 2.2  **k**-space in 2D and 3D. Minimum separation between allowable wavevectors is $\frac{2\pi}{L}$. **k**-space spherical 'volume' (circular area for 2D) is depicted in the figure.

The resulting density of states for a bulk material becomes:

$$D_{3D,\text{Debye}}(\omega) = \frac{\omega^2}{2\pi^2 v_{g,\text{ave}}^3}, \text{ for } \omega < \omega_D = v_{g,ave}(6\pi^2 \eta_a)^{1/3}, \qquad (2.22)$$

where $v_{g,\text{ave}}$ is an appropriately averaged phonon velocity among the acoustic polarizations and $\eta_a$ is the number of unit cells per unit volume of real space. For now, we will refrain from making this approximation.

## 2.4   Electron Density of States

Electron states are similarly restricted, as shown in Section 1.5 for the particle-in-box problem. The result, Eq. (1.37), in combination with the parabolic dispersion of Eq. (1.34) provides us with analogues to the foregoing analysis of phonons.

The electron density of states is almost always expressed per unit energy $E$ as $D(E)$ (this convention allows us to distinguish it from the phonon density of states, which is usually described in terms of frequency $\omega$, *cf.* Eq. (2.21)). Given the quantum relation $E = \hbar\omega$ between energy and frequency, the density of states per unit energy is closely related to that given above for phonon density of states per unit frequency. Accordingly, the electronic density of states can be expressed as $D(E) = 2D(\omega)/\hbar$, where the factor of 2 accounts for spin degeneracy. The resulting expressions in each dimensionality follow:

$$D_{1D}(E) = \frac{2m_e}{\pi\hbar^2 k(E)}, \qquad (2.23)$$

$$D_{2D}(E) = \frac{m_e}{\pi\hbar^2}, \qquad (2.24)$$

$$D_{3D}(E) = \frac{m_e k(E)}{\pi^2\hbar^2}, \qquad (2.25)$$

where $m_e$ is the electron rest mass, and we have made use of the momentum relation $m_e v_g = \hbar k$.

The same result can be derived by integrating over **k**-space using Dirac delta functions for allowable states (Lundstrom, 2009):

$$D(E) = \frac{1}{L^d} \sum_{k'} \delta\left[E(k) - E(k')\right],$$

$$= \frac{2}{(2\pi)^d} \int_{k'} \delta\left(\frac{\hbar^2 k^2}{2m_e} - \frac{\hbar^2 k'^2}{2m_e}\right) d\vec{k'}$$

$$= \frac{2m_e}{\pi\hbar\sqrt{2m_e E}} \quad (1D)$$

$$= \frac{m}{\pi\hbar^2} \quad (2D)$$

$$= \frac{m_e\sqrt{2m_e E}}{\pi^2\hbar^3} \quad (3D), \qquad (2.26)$$

where in this set of equations we have made the parabolic band approximation for $k(E)$, namely:

$$E = \frac{\hbar^2 k^2}{2m_e} \quad \text{or} \quad k(E) = \frac{\sqrt{2m_e E}}{\hbar}. \tag{2.27}$$

Figure 2.3 shows a parabolic electron band with two ranges of **k**-space. The allowed states are equally distributed in **k**-space, but corresponding mapping of energies from allowed **k**-states shows a higher density of states at low energies (and low wavevectors), where the band is flattest.

A schematic of the resulting electron density of states appears in Fig. 2.4. The curves for confined structures (quantum wells, wires, and dots) contain multiple 'bands' that build upon each other and generally follow the overall $\sqrt{E}$ trend of the curve for a bulk conductor.

For greater depth, the reader is referred to the nanoHUB's 'CNTbands' tool (Seol *et al.*, 2011a), which calculates the geometry, band structure, and density of states of single-walled carbon nanotubes. As an example, Fig. 2.5 shows results for a (12, 12) CNT.

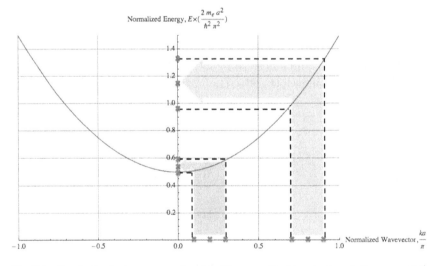

Fig. 2.3   Parabolic electron energy band (with normalized band edge at $E_{c,\text{norm}} = 0.5$) and corresponding allowable $k$-states at low and high wavevectors.

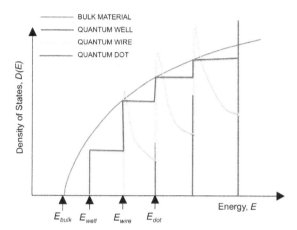

Fig. 2.4 Electron density of states for 0D (quantum dot), 1D (quantum wire), 2D (quantum well), and 3D (bulk) materials. Bulk material density of states follows a $\sqrt{E}$ dependence, whereas confined material densities of states present discontinuities due to multiple band-folding from confined dimensions.

## 2.5 Example: Derivation of Planck's Law

This section provides a brief derivation of Planck's law of blackbody radiation from basic statistical principles, as an example of a 'boson' thermal field. For more information, the reader is referred to the textbook by Rybicki and Lightman (2008). The reader might also find interest in the historical development of early research in radiation physics as surveyed by Barr (1960).

### 2.5.1 *Photon Gas in a Box*

First, consider a cubic box with each side of length $L$ filled with electromagnetic (EM) radiation (a so-called 'photon gas') that forms standing waves whose allowable wavelengths are restricted by the size of the box. We will assume that the waves do not interact and therefore can be separated into the three orthogonal Cartesian directions such that the allowable wavelengths are:

$$\lambda_i = \frac{2L}{n_i}, \tag{2.28}$$

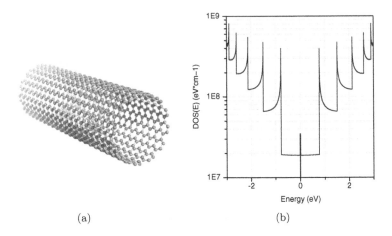

(a)                                                    (b)

Fig. 2.5  Geometry and density of states of a (12,12) single-walled carbon nanotube (SWCNT). Refer to https://nanohub.org/resources/cntbands-ext for the online tool.

where $n_i$ is an integer greater than zero, and $i$ represents one of the three Cartesian directions–$x, y$, or $z$.

From quantum mechanics, the energy of a given mode (i.e., an allowable set $n_x, n_y, n_z$) can be expressed as:

$$E(N) = \left(N + \frac{1}{2}\right) \frac{hc}{2L} \sqrt{n_x^2 + n_y^2 + n_z^2}, \qquad (2.29)$$

where $h$ is Planck's constant ($6.626 \times 10^{-34}$J s). The number $N$ represents the number of such modes, or photons, with the set of quantum numbers $\{n_x, n_y, n_z\}$. Importantly, unlike electrons, an unlimited number of modes, or photons, of a given energy can exist; thus, photons are governed by Bose-Einstein statistics, for which the average $\langle N \rangle = f_{BE}^o$ according to Eq. (2.9).

### 2.5.2  *Statistical Mechanics of the Photon Gas*

To derive the energy density in this photon gas, we first need to know the relative probability with which a given energy state $E(N)$ is occupied at a given temperature. Here, we turn to statistical mechanics, which reveals this probability as:

$$P_N = \frac{\exp(-\beta E(N))}{Z(\beta)}, \qquad (2.30)$$

where $\beta$ is the inverse of thermal energy, or $\beta = (k_B T)^{-1}$, and $Z(\beta)$ is the partition function that normalizes the probability as:

$$Z(\beta) = \sum_{N=0}^{\infty} \exp(-\beta E(N)) = \frac{1}{1 - \exp(-\beta\varepsilon)}, \qquad (2.31)$$

where $\varepsilon = \frac{hc}{2L}\sqrt{n_x^2 + n_y^2 + n_z^2} = \frac{hc}{\lambda}$ is the energy of a single photon, and the latter equality derives from the relationship between the wavelength $\lambda$ and the $n_i$ indices of the EM waves in the box. This wavelength is related to the speed of light $c$ and frequency $\nu$ through the familiar relation

$$\frac{c}{\lambda} = \nu \Rightarrow \varepsilon = h\nu. \qquad (2.32)$$

Again from statistical mechanics (and specifically Bose-Einstein statistics), the average energy within a given mode can be expressed as

$$\langle E(N) \rangle = -\frac{d \ln Z}{d\beta} = \frac{\varepsilon}{\exp(\beta\varepsilon) - 1}, \qquad (2.33)$$

where the zero point energy is neglected.

### 2.5.3 *Energy Density of the Photon Gas*

Now that we have an expression for the average energy of a given mode, we can sum (integrate) over all modes to find the total specific energy within the photon gas. This energy can be expressed as an integral over all energies:

$$u = \int_0^{\infty} \langle E \rangle D(\varepsilon) d\varepsilon$$

$$= \int_0^{\infty} \frac{\varepsilon}{\exp(\beta\varepsilon) - 1} D(\varepsilon) d\varepsilon, \qquad (2.34)$$

where $D(\varepsilon)$ is the density of states that gives the number of allowed modes per unit volume and per unit energy within an interval between $\varepsilon$ and $\varepsilon + d\varepsilon$. This function can be derived from the allowable wavelengths and $n$ indices as:

$$D(\varepsilon)d\varepsilon = \frac{8\pi}{h^3 c^3}\varepsilon^2 d\varepsilon. \qquad (2.35)$$

The total energy per unit volume (or 'specific' energy) can now be expressed as

$$u = \int_0^\infty \frac{8\pi}{h^3 c^3} \frac{\varepsilon^3}{\exp(\beta\varepsilon) - 1} d\varepsilon, \tag{2.36}$$

where the integrand is the spectral energy density $u'$. This function can be expressed in terms of an energy, wavelength, or frequency spectral basis through the relation $\varepsilon = hc/\lambda$ such that different forms of $u'$ are commonly used. However, they are each integrands in expressions that are used to calculate the overall energy density as:

$$\frac{U}{L^3} = u(T) = \int_0^\infty u'(\varepsilon, T) d\varepsilon = \int_0^\infty u'(\lambda, T) d\lambda = \int_0^\infty u'(\nu, T) d\nu. \tag{2.37}$$

The corresponding expressions for spectral energy density follow:

$$u'(\varepsilon, T) = \frac{8\pi}{h^3 c^3} \frac{\varepsilon^3}{\exp\left(\frac{\varepsilon}{k_B T}\right) - 1}, \tag{2.38}$$

$$u'(\lambda, T) = \frac{8\pi hc}{\lambda^5} \frac{1}{\exp\left(\frac{hc}{\lambda k_B T}\right) - 1}, \tag{2.39}$$

$$u'(\nu, T) = \frac{8\pi h\nu^3}{c^3} \frac{1}{\exp\left(\frac{h\nu}{k_B T}\right) - 1}, \tag{2.40}$$

$$u'(\omega, T) = \frac{(\hbar\omega)^3}{\pi^2 \hbar^2 c^3} \frac{1}{\exp\left(\frac{\hbar\omega}{k_B T}\right) - 1}. \tag{2.41}$$

### 2.5.4  Blackbody Emission Intensity

Now assume that a small hole is cut into the box as shown in Fig. 2.6. All radiation emanating from this hole will be moving at the speed of light $c$. Also, the radiation will be uniformly distributed throughout the hemisphere of solid angles ($2\pi$ steradians), and one half of the energy will be oriented such that it can move outward through the hole.

The spectral radiation intensity is defined as the rate of energy emitted per unit area per unit solid angle and per unit wavelength. The rate of energy emitted per area is simply the product of the energy density derived

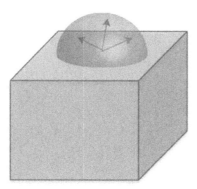

Fig. 2.6    Blackbody emission from a small hole in a box.

above and the speed of light (i.e., the distance swept by a ray per unit of time). Therefore, the spectral intensity becomes:

$$I(\lambda, T) = \frac{1}{2} \left[ \frac{u'(\lambda, T)c}{2\pi} \right] = \frac{2hc^2}{\lambda^5} \frac{1}{\exp\left(\frac{hc}{\lambda k_B T}\right) - 1}. \tag{2.42}$$

Similarly, the spectral intensity (per unit frequency $\nu$ instead of wavelength) is:

$$I(\nu, T) = \frac{1}{2} \left[ \frac{u'(\nu, T)c}{2\pi} \right] = \frac{2h\nu^3}{c^2} \frac{1}{\exp\left(\frac{h\nu}{k_B T}\right) - 1}. \tag{2.43}$$

And finally, the intensity per unit angular frequency $\omega = 2\pi\nu$ is:

$$I(\omega, T) = \frac{1}{2} \left[ \frac{u'(\omega, T)c}{2\pi} \right] = \frac{\hbar\omega^3}{4\pi^3 c^2} \frac{1}{\exp\left(\frac{\hbar\omega}{k_B T}\right) - 1}. \tag{2.44}$$

This distribution is plotted for different temperatures in Fig. 2.7.

The foregoing analysis of electromagnetic transport with photons highlights the convergence of statistical distributions and energy states. The results emphasize particularly well the spectral behavior of transport and its interrelationship with temperature. Subsequent chapters will demonstrate similar concepts for phonons and electrons.

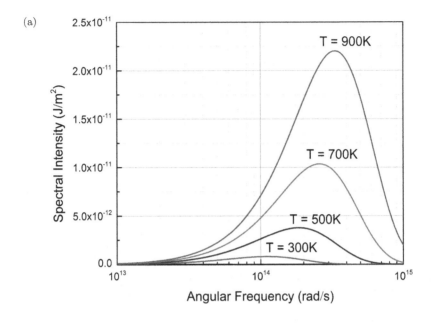

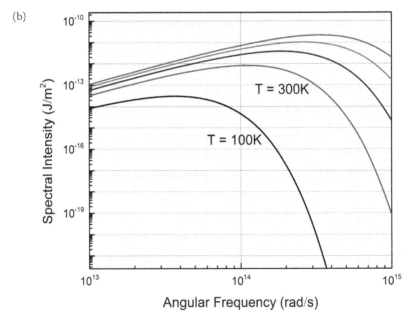

Fig. 2.7   Spectral intensity (per unit angular frequency $\omega$) as a function of angular frequency at different temperatures. The frequency at maximum spectral intensity increases with increasing temperature, according to Wien's displacement law (Modest, 2003).

# Example Problems

---

**Problem 2.1: Getting a feel for the numbers (note: this problem has been adapted from Kaviany (2008))**

(a) The maximum energies of acoustic and optical phonons in graphene are 0.16 eV and 0.21 eV respectively. Determine $f^o_{BE}$ at $T = 300$ and 3000 K, for these two energies.

(b) The Fermi energy of aluminum is 11.7 eV. Assuming that the chemical potential is equal to the Fermi energy, determine $f^o_{FD}$ for $E = 1$, 11.5 and 20 eV. Calculate the occupation numbers at $T = 1$ and 3000 K.

(c) The average thermal speed of monoatomic gas molecules is given by $\sqrt{8k_B T/(\pi m)}$. Determine the average speed, kinetic energy and the Maxwell-Boltzmann energy distribution function $f^o_{MB}$ (at the average energy) for argon gas at $T = 300$ K.

(d) The surface temperature of the sun can be approximated to be about 5700 K. Determine $f^o_{BE}$ for photons emitted from the sun at $\lambda = 100$ nm (UV), $\lambda = 600$ nm (visible) and $\lambda = 900$ nm (IR).

*Solution*

(a) The Bose-Einstein distribution is given by:

$$f^o_{BE} = \frac{1}{\exp(E/k_B T) - 1}. \qquad (2.45)$$

At $T = 300$ K, $k_B T = 0.026$ eV. Thus, $f^o_{BE} = 0.002$ for $E = 0.16$ eV and $f^o_{BE} = 0.0003$ for $E = 0.21$ eV. At $T = 3000$ K, $k_B T = 0.258$ eV. Thus, $f^o_{BE} = 1.164$ for $E = 0.16$ eV and $f^o_{BE} = 0.796$ for $E = 0.21$ eV. The occupation numbers increase with increasing temperature (see Fig. 2.8a). Also note that $f^o_{BE}$ can be greater than 1 because the Pauli exclusion principle does not apply for bosons.

(b) The Fermi-Dirac distribution is given by:

$$f^o_{FD} = \frac{1}{\exp((E - \mu)/k_B T) + 1}. \qquad (2.46)$$

At $T = 1$ K, $k_B T = 8.63 \times 10^{-5}$ eV. Thus $f^o_{FD} = 1$ for $E = 1$ eV, $f^o_{FD} = 1$ for $E = 11.5$ eV and $f^o_{FD} = 0$ for $E = 20$ eV.

At $T = 3000$ K, $k_B T = 0.258$ eV. Thus $f_{FD}^o = 1$ for $E = 1$ eV, $f_{FD}^o = 0.68$ for $E = 11.5$ eV and $f_{FD}^o = 0$ for $E = 20$ eV. The Fermi-Dirac distribution changes from 1 to 0 in a small energy window (of the order of $k_B T$) around the electrochemical potential (see Fig. 2.8b).

(c) From the given expression, the average thermal speed of argon atoms ($m = 40$ amu. $= 6.64 \times 10^{-26}$ kg) at $T = 300$ K is 398.8 m/s. Thus the average kinetic energy is:

$$E = \frac{1}{2} m_{Ar} v^2 = 0.033 \text{ eV}. \qquad (2.47)$$

The Maxwell-Boltzmann distribution is given by $f_{MB}^o = \exp(-E/k_B T) = 0.28$.

(d) For a given wavelength $\lambda$, the energy of a photon is $E = hc/\lambda$. Thus $E = 12.42$ eV for $\lambda = 100$ nm, $E = 2.07$ eV for $\lambda = 600$ nm and $E = 1.38$ eV for $\lambda = 900$ nm. Also $k_B T = 0.49$ eV at $T = 5700$ K. Thus $f_{BE}^o = 9.82 \times 10^{-12}$ for $\lambda = 100$ nm, $f_{BE}^o = 0.0148$ for $\lambda = 600$ nm and $f_{BE}^o = 0.063$ for $\lambda = 900$ nm.

---

**Problem 2.2: Working with the Bose-Einstein distribution function**

(a) The energy levels of a quantum harmonic oscillator are given by:

$$E_n = \left( n + \frac{1}{2} \right) \hbar \omega,$$

where $n = 0, 1, 2 \ldots$. Obtain an expression for the partition function $\Xi$ (you will need to sum an infinite geometric series) defined by ($\mu = 0$ for phonons):

$$\Xi = \sum_n \exp(-\beta E_n).$$

(b) Obtain an expression for the average energy $\langle E \rangle$ defined by:

$$\langle E \rangle = -\frac{\partial \ln \Xi}{\partial \beta}.$$

Show that the average energy of the mode with frequency $\omega$ can be written as $\hbar \omega \left( f_{BE}^o + \frac{1}{2} \right)$, where $f_{BE}^o$ denotes the Bose-Einstein distribution function.

*Solution*

(a) The partition function $\Xi$ is given by:

$$\Xi = \sum_n \exp\left(-\beta E_n\right)$$

$$= \sum_{n=0}^{n=\infty} \exp\left(-\beta\left(n + 1/2\right)\hbar\omega\right)$$

$$= \frac{\exp(-\beta\hbar\omega/2)}{1 - \exp(-\beta\hbar\omega)}. \tag{2.48}$$

(b) The average energy $\langle E \rangle$ is given by:

$$\langle E \rangle = -\frac{\partial ln\Xi}{\partial\beta}$$

$$= -\frac{\partial}{\partial\beta}\left(-\frac{\beta\hbar\omega}{2} - \ln(1 - \exp(-\beta\hbar\omega))\right)$$

$$= \frac{\hbar\omega}{2} + \frac{\hbar\omega\exp(-\beta\hbar\omega)}{1 - \exp(-\beta\hbar\omega)}$$

$$= \hbar\omega\left(\frac{1}{\exp(\beta\hbar\omega) - 1} + \frac{1}{2}\right)$$

$$= \hbar\omega\left(f_{BE}^o + \frac{1}{2}\right). \tag{2.49}$$

---

**Problem 2.3: Phonon DOS in graphene**

The dispersion relation for graphene (excluding optical branches), which is a two-dimensional material, is shown in Fig. 2.9. Graphene has three acoustic branches commonly known as the LA, TA and ZA modes. The LA and TA modes can be approximated by a linear dispersion relation while the ZA mode, which represents out-of-plane vibrations, is more closely represented by a quadratic dispersion relation near the Brillouin zone center (see Appendix). Obtain an expression for the phonon density of states $D(\omega)$ at three different frequencies $\omega_1$, $\omega_2$, and $\omega_3$ as indicated in Fig. 2.9.

*Solution*

The DOS for the LA mode is given by:

$$D_{LA}(\omega) = \frac{1}{L^2} \frac{dN}{dK} \frac{dK}{d\omega}$$

$$= \frac{1}{v_{g1} L^2} \frac{L^2 K}{2\pi} \qquad \left( N = \frac{\pi K^2}{(2\pi/L)^2}, \frac{d\omega}{dK} = v_{g1} \right)$$

$$= \frac{\omega}{2\pi v_{g1}^2}. \tag{2.50}$$

Similarly, the DOS for the linear TA mode is given by:

$$D_{TA}(\omega) = \frac{\omega}{2\pi v_{g2}^2}. \tag{2.51}$$

The DOS for the quadratic ZA mode is given by:

$$D_{ZA}(\omega) = \frac{1}{L^2} \frac{dN}{dK} \frac{dK}{d\omega}$$

$$= \frac{1}{2cK L^2} \frac{L^2 K}{2\pi} \qquad \left( N = \frac{\pi K^2}{(2\pi/L)^2}, \frac{d\omega}{dK} = 2cK \right)$$

$$= \frac{1}{4\pi c}. \tag{2.52}$$

All three modes are present at $\omega_1$. Thus,

$$D_{\text{tot}}(\omega_1) = D_{LA}(\omega_1) + D_{TA}(\omega_1) + D_{ZA}(\omega_1)$$

$$= \frac{\omega_1}{2\pi v_{g1}^2} + \frac{\omega_1}{2\pi v_{g2}^2} + \frac{1}{4\pi c}. \tag{2.53}$$

Only the LA and TA modes are active at $\omega_2$:

$$D_{\text{tot}}(\omega_2) = D_{LA}(\omega_2) + D_{TA}(\omega_2)$$

$$= \frac{\omega_2}{2\pi v_{g1}^2} + \frac{\omega_2}{2\pi v_{g2}^2}. \tag{2.54}$$

Only the LA mode is active at $\omega_3$:

$$D_{\text{tot}}(\omega_3) = D_{LA}(\omega_3)$$

$$= \frac{\omega_3}{2\pi v_{g1}^2}. \tag{2.55}$$

**Problem 2.4: Phonon frequency at maximum intensity**

Wien's displacement law for electromagnetic radiation relates the photon wavelength at which the energy distribution is maximum to the temperature. In this problem we derive a similar relation for phonons, except in frequency space. Assume a 3D material with a single phonon branch having a constant group velocity $v_g$ (Debye approximation).

• Show that the dominant phonon frequency (frequency at which the spectral energy distribution is maximum) $\omega_{max}$ as a function of temperature $T$ is given by $\hbar\omega_{max} = Ck_BT$ where $C$ is a constant of proportionality. Neglect the zero-point energy in your analysis.

• Use the online Chapter 2 CDF tool[1] to observe the spectral phonon energy distribution as a function of temperature.

• Verify the relation you obtained for the maximum phonon frequency by tabulating the maximum points in the curve for a few different temperatures. Also obtain the constant $C$.

2.5.4.1 *Solution*

Under the Debye approximation, the density of states $D(\omega)$ is given by:

$$D(\omega) = \frac{\omega^2}{2\pi^2 v_g^3}. \tag{2.56}$$

The spectral energy density $u(\omega, T)$ is then given by:

$$u(\omega, T) = \underbrace{\hbar\omega}_{\text{energy}} \underbrace{\frac{\omega^2}{2\pi^2 v_g^3}}_{\text{DOS}} \underbrace{\frac{1}{\exp(\hbar\omega/k_BT) - 1}}_{\text{occupation}}$$

$$= \frac{\hbar}{2\pi^2 v_g^3} \frac{\omega^3}{\exp(\hbar\omega/k_BT) - 1}. \tag{2.57}$$

At a given temperature, spectral energy density is a maximum when $\frac{\partial u}{\partial \omega} = 0$. Thus,

$$3\omega^2(\exp(\hbar\omega/k_BT) - 1) - \omega^3 \exp(\hbar\omega/k_BT)\frac{\hbar}{k_BT} = 0. \tag{2.58}$$

[1]See http://nanohub.org/groups/cdf_tools_thermal_energy_course/wiki

Defining $x^* = \hbar\omega/k_B T$, we arrive at the following implicit equation for $x^*$.

$$3(1 - \exp(-x^*)) = x^*. \qquad (2.59)$$

- The above equation can be solved numerically (using WolframAlpha[2] for example) to obtain $x^* = 2.82$. Thus $\hbar\omega_{max} = 2.82\, k_B T$. The frequency at which the spectral energy distribution is a maximum increases linearly with temperature. In other words, the peak wavelength is inversely proportional to temperature.
- Figure 2.10 shows snapshots from the online Chapter 2 CDF tool where the spectral energy distribution is plotted for three different temperatures. The peak of the curves moves to the right for increasing temperature.
- Use the online tool to tabulate $\omega_{max}$ for a few different temperatures and confirm the linear relationship between $\omega_{max}$ and $T$.

---

[2]See http://www.wolframalpha.com/input/?1=solve+3(1-exp(-x))+%3D+x

(a)

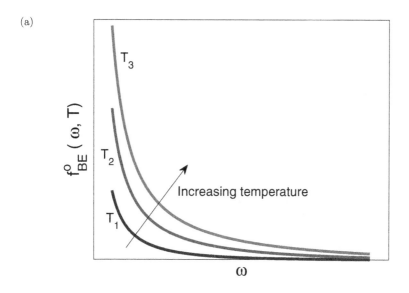

(b)

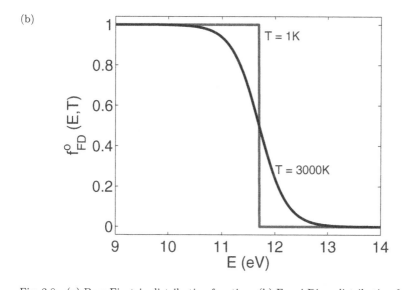

Fig. 2.8 (a) Bose-Einstein distribution function. (b) Fermi-Dirac distribution function.

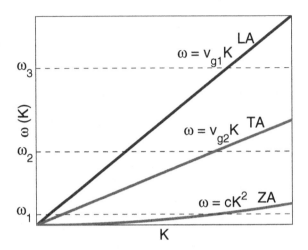

Fig. 2.9    Graphene dispersion relation.

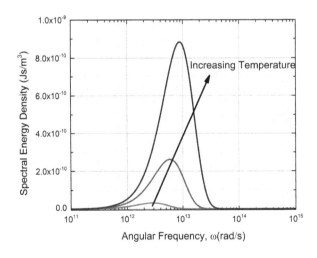

Fig. 2.10    Spectral energy distribution for three different temperatures.

Chapter 3

# Basic Thermal Properties

## 3.1 Introduction

The energy density analysis discussed in Section 2.5.3 provided important insights about the inter-relationship between carrier energy and carrier statistics for photons. A similar analysis for phonons and electrons (and their respective energy levels and statistics) will provide the basis for the property known as internal energy. This property should be familiar to those who have studied classical or statistical thermodynamics, as should the related quantity called specific heat. For a given collection of carriers, knowledge of its internal energy and dependence on temperature (which derives from its statistics) allows explicit calculation of volumetric specific heat as:

$$c_v = \frac{\partial u(T)}{\partial T},$$ (3.1)

where the normalizing quantity (i.e., the 'amount' of the ensemble by which $u(T)$ is normalized) can be either volume or mass. In Eq. (2.37) from the previous chapter, volume is the normalizing quantity for $u(T)$.

The specific heat quantifies the ability of a set of energy carriers to store thermal energy relative to the temperature rise required to store this energy. At the same time, these carriers can move within a material or control volume and while doing so transport thermal energy. Consequently, the average speed with which the carriers move combined with the amount of energy that they carry provides the foundation of the important thermal transport property known as thermal conductivity. We develop these concepts in the present chapter for the carriers of most interest here—phonons and electrons.

## 3.2    Specific Heat

The most general expression for the extensive (i.e., not specific, or intensive) internal energy $U$ is:

$$U = \sum_{\mathbf{k}} \sum_{p} E_{i,p}(\mathbf{k}) f_i^o \left[ E_{i,p}(\mathbf{k}), T \right], \qquad (3.2)$$

where we have neglected zero-point energy. The foregoing equation is the summation form of the integral expression (Eq. (2.34)) used in the derivation of Planck's law for photons, in which the concept of density of states was employed somewhat obsequiously to convert sums to integrals. This duality between summative and integral forms of quantities will persist throughout subsequent analysis both because of various preferences that have evolved in different communities of theorists and because sometimes the summative form is more analytically convenient than the integral form and *vice versa*, depending on context. Here, we seek first to relieve some of the common confusion associated with the dual forms.

We first recognize that $\mathbf{k}$-space summation is often cumbersome. A general conversion from summation to integration of a function $F$ in $\mathbf{k}$-space is:

$$\lim_{L \to \infty} \frac{1}{L^d} \sum_{\mathbf{k}} F(\mathbf{k}) = \int F(\mathbf{k}) \frac{d\mathbf{k}}{(2\pi)^d}, \qquad (3.3)$$

where $F(\mathbf{k})$ is a generic function in a $\mathbf{k}$-space of dimension $d$ (which matches the real-space dimension of a given problem). This conversion derives from the fact that each allowable states $\mathbf{k}$-space volume is $(2\pi/L)^d$. Applied to Eq. (3.2), the internal energy can be expressed in integral form as:

$$u = \frac{U}{L^d} = \sum_{p} \int \frac{E_{i,p}(\mathbf{k}) f_i^o \left[ E_{i,p}(\mathbf{k}), T \right]}{(2\pi)^d} d\mathbf{k}. \qquad (3.4)$$

And then using Eq. (3.1), specific heat can be expressed as:

$$c_v = \frac{\partial u}{\partial T} = \sum_{p} \int \frac{E_{i,p}(\mathbf{k})}{(2\pi)^d} \frac{\partial f_i^o}{\partial T} d\mathbf{k}, \qquad (3.5)$$

where

$$\frac{\partial f_i^o}{\partial T} = (f_i^o)^2 e^{(E_i - \mu)/k_B T} \left( \frac{E_i - \mu}{k_B T^2} \right). \qquad (3.6)$$

For many (most, really) problems, the $d$-dimensional **k**-space integral remains cumbersome, and therefore, the concept of density of states is used to convert the multi-dimensional integral to a single dimension (either energy or frequency). We apply this process to different carrier types in the following subsections.

### 3.2.1 *Acoustic Phonon Specific Heat*

The **k**-space integral in Eq. (3.5) explicitly calls for knowledge of the energy-wavevector relation $(E_{i,p}(\mathbf{k}))$. For phonons, this relation is typically expressed in terms of frequency and wavevector through the dispersion relation. In Chapter 1, we derived the dispersion relation for an acoustic phonon branch in the idealized one-dimensional atomic chain (Eq. (1.22)). In higher dimensions, an atom can move in more than one direction, as shown in Fig. 3.1. These extra dimensions create additional phonon 'branches' whose dispersion is generally similar to the longitudinal branch (see Fig. 1.13), except that the effective spring constant $g$ differs, resulting in a different maximum phonon frequency at the edge of the Brillouin zone. Further, three-dimensional crystals with appropriate lattice symmetry relative to the propagation direction of interest can exhibit *degeneracy* such that the two transverse branches have identical dispersion relations.

The curvature of the acoustic phonon dispersion relation and associated factors that cause real materials to deviate from the ideal sine function of Eq. (1.22) has motivated the use of a simplified dispersion model. The most prominent among these is the Debye approximation (Debye, 1912), which approximates the sine function as a line through the $\omega$-$K$ origin:

$$\omega(K) \approx v_{g,\text{ave}} K. \tag{3.7}$$

This approximation, however, cannot be applied blindly because doing so would fail to account for the finite number of allowable phonon states as discussed in Section 2.3. The Debye frequency represents the maximum allowable frequency such that the number of states $(N)$ in a given branch matches the number of allowed wavevectors.

To determine the number of independent wavevectors in a Brillouin zone, the following statement from Ziman (1972, p. 25) is crucial:

> [T]here are exactly as many allowed wave-vectors in a Brillouin zone as there are unit cells in a block of crystal.

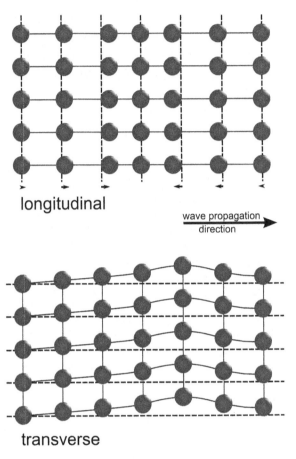

**longitudinal**

wave propagation direction

**transverse**

Fig. 3.1 Longitudinal and transverse phonon branches. Atoms vibrate along the wave propagation direction in the longitudinal mode. In the transverse mode, atoms vibrate perpendicular to the wave propagation direction.

The number of allowed states in each dimensionality as a function of the magnitude of the wavevector $K$ was given previously (Eq. (2.13)–(2.15)). Setting the number of states to equal the number of unit cells in the crystal, the Debye wavevectors are:

$$K_{D,1D} = \pi\eta_a, \tag{3.8}$$

$$K_{D,2D} = (4\pi\eta_a)^{1/2}, \tag{3.9}$$

$$K_{D,3D} = (6\pi^2\eta_a)^{1/3}, \tag{3.10}$$

where $\eta_a$ is the number of unit cells per unit 'volume' of the given dimensionality (i.e., length for 1D, area for 2D, and true volume for 3D). Substituting the Debye dispersion relation gives the Debye frequency:

$$\omega_{D,1D} = v_{g,\text{ave}}\pi\eta_a, \tag{3.11}$$

$$\omega_{D,2D} = v_{g,\text{ave}}(4\pi\eta_a)^{1/2}, \tag{3.12}$$

$$\omega_{D,3D} = v_{g,\text{ave}}(6\pi^2\eta_a)^{1/3}. \tag{3.13}$$

A common tabulated representation of the Debye frequency for bulk (3D) materials is the Debye temperature:

$$\theta_D = \frac{\hbar\omega_{D,3D}}{k_B} = \frac{\hbar v_{g,\text{ave}}(6\pi^2\eta_a)^{1/3}}{k_B}. \tag{3.14}$$

One subtle point of clarification is important here, because the foregoing Debye quantities are often expressed in terms of the atomic density instead of the unit cell density. The former is substantially easier to calculate because it can be derived easily from knowledge of a crystal's mass density and constituent atomic mass(es). Of course, the two are equivalent for crystals with one atom per unit cell (i.e., with no basis atoms). In contrast, as described in Section 1.6 the presence of basis atoms produces entirely new phonon branches (cf., the optical branch in the 1D diatomic chain example). If our intention is to approximate *both* acoustic and optical branches with the linear-dispersion Debye approximation, then we would replace the unit cell density $\eta_a$ in the foregoing equations with $n \times \eta_a$, where $n$ represents the number of atoms per unit cell.

This approximation may be reasonable with the understanding that the Debye wavevector will extend substantially beyond the edge of the Brillouin zone boundary, as shown in Fig. 3.2, to include the extension of the optical branch into the 'second' Brillouin zone. However, as we will see in the subsequent section, an entirely different model for specific heat is often most appropriate for optical phonons, in which case the unit cell density alone should be used in calculating Debye metrics. For readers seeking further details, Ashcroft and Mermin (1976, pp. 462–463) clarify this issue particularly well.

The general results for phonon density of states in frequency space (Eqs. (2.19)–(2.21)) become, under the Debye approximation:

$$D_{D,1D}(\omega) = \frac{1}{v_{g,\text{ave}}\pi}, \tag{3.15}$$

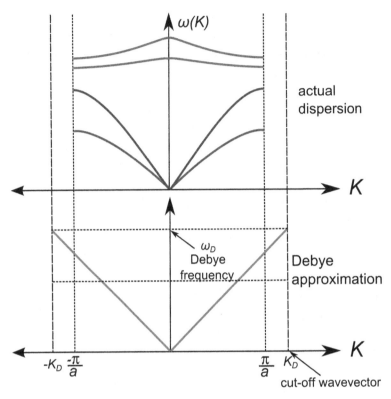

Fig. 3.2  Debye's linear approximation to the phonon dispersion. The Debye cutoff wavevector, $K_D$, is chosen such that it contains allowed wavevectors equalling the number of ions in the crystal. Debye quantities such as the Debye cutoff wavevector, $K_D$, and the associated Debye cutoff frequency, $\omega_D$, are depicted.

$$D_{D,2D}(\omega) = \frac{\omega}{2\pi v_{g,\text{ave}}^2}, \qquad (3.16)$$

$$D_{D,3D}(\omega) = \frac{\omega^2}{2\pi^2 v_{g,\text{ave}}^3}, \qquad (3.17)$$

where these results are applicable for $\omega < \omega_D$. For most of the foregoing Debye quantities, the average group velocity plays an important role. Often, the long-wavelength velocity $\left[\frac{\partial \omega}{\partial K}\right]_{K=0}$ is used. This approximation is most suitable for low-temperature conditions in which the phonon population is dominated by low frequencies. However, even in such cases, the single value of $v_{g,\text{ave}}$ implies **k**-space symmetry in the applicable directions. Further, in some prior work, a single group velocity is defined as an amalgam of the transverse and longitudinal branches. Instead of enumerating all of the possible representations for group velocity here, the reader is cautioned to scrutinize the definition of group velocity for any work that employs the Debye approximation.

Given the foregoing assumptions (and particularly re-emphasizing the k-space symmetry), we can calculate the canonical Debye specific heat for acoustic phonons in 3D (bulk) materials from the general expression (Eq. (3.5)):

$$
c_{v,D,3D} = \sum_p \int \frac{E_{i,p}(K)}{(2\pi)^d} \frac{\partial f_i^o}{\partial T} d\mathbf{k} \overset{3D}{=} \sum_p \int \frac{\hbar v_{g,\text{ave}} K}{(2\pi)^3} \frac{\partial f_i^o}{\partial T} d\mathbf{k}
$$

$$
= \sum_p \int_0^{K_D} \frac{\hbar v_{g,\text{ave}} K}{(2\pi)^3} \frac{\partial f_{BE}^o}{\partial T} 4\pi K^2 dK
$$

$$
= \sum_p \int_0^{\omega_D} \hbar\omega \frac{\partial f_{BE}^o}{\partial T} D_{D,3D}(\omega) d\omega
$$

$$
= \sum_p \int_0^{\omega_D} \frac{\hbar\omega^3}{2\pi^2 v_{g,\text{ave}}^3} \frac{\partial f_{BE}^o}{\partial T} d\omega
$$

$$
= 9\eta_a k_B \left(\frac{T}{\theta_D}\right)^3 \int_0^{\theta_D/T} \frac{x^4 e^x dx}{(e^x - 1)^2}, \tag{3.18}
$$

where $x \equiv \frac{\hbar\omega}{k_B T}$, and the last equality derives from (a) the definition of the temperature derivative of the distribution function (Eq. (3.6)), (b) the definition of group velocity in terms of unit cell density and Debye temperature (Eq. (3.14)) and (c) the assumption that the three acoustic phonon branches can be combined through a single Debye group velocity (thus eliminating the branch summation by multiplying the integral by a factor of 3). This latter approximation should be taken with caution; again, the definition of the group velocity must be appropriate to the assumptions invoked.

The final Debye specific heat expression in Eq. (3.18) is a much-celebrated result, despite the fact that the integral is not generally reducible to an analytic expression. Often, the extreme temperature limits can be used to determine limiting expressions for low and high temperatures *relative to the Debye temperature*. For the low-temperature limit:

$$
c_{v,D,3D\text{low}} \approx 234\eta_a k_B \left(\frac{T}{\theta_D}\right)^3 \quad (T \ll \theta_D). \tag{3.19}
$$

This result indicates that, at low temperatures, the acoustic phonon specific heat increases as $T^3$ for a 3D material, and we will find below that the factor of 3 derives from the dimensionality of the material (e.g., the temperature dependence for 2D materials is $T^2$). Conversely, for very high temperatures (relative to $\theta_D$), the result becomes independent of temperature:

$$c_{v,D,\text{3Dhigh}} \approx 3\eta_a k_B \quad (T \gg \theta_D). \tag{3.20}$$

The temperature-independence of the latter result is caused by the capping of energy states imposed by the maximum phonon energy associated with $\theta_D$. In effect, any increases in temperature must be accommodated by increasing the phonon populations of states that are already well occupied at or below $\theta_D$, as opposed to the situation at very low temperatures for which empty states can be filled when temperature increases. The high-temperature result is called the "Law of Dulong and Petit" (Ashcroft and Mermin, 1976).

### 3.2.2 Optical Phonon Specific Heat

Clearly, the Debye specific heat model is well suited for phonon branches that exhibit a linear-like behavior through the origin of the dispersion curve (e.g., acoustic branches with shapes like quarter sine waves). However, this model is dubious for optical phonons, which exhibit relatively flat dispersion curves that intercept the frequency (energy) axis at non-zero values (see Fig. 1.18). Einstein (1906) proposed a general model for phonon specific heat that assumes such flat dispersion behavior by assigning a single phonon frequency to each branch. While the intention was to apply this model for all branches, later developments, such as the Debye model (Debye, 1912) described above, revealed clearly that this model is best applied to optical phonon branches only.

The derivation of the Einstein model for specific heat generally follows that of Section 3.2.1. It differs in the assumption that all phonons in the branch of interest oscillate at a single frequency $\omega_E$, leading to the definition of the Einstein temperature $\theta_E$:

$$\theta_E = \frac{\hbar \omega_E}{k_B}. \tag{3.21}$$

For convenience, we retain the summation form throughout the derivation. Using this dispersion relation (i.e., $\omega = \text{constant} = \omega_E$) in the general integral expression for internal energy $U$ (Eq. (3.2)) results in:

$$U = \sum_{\mathbf{k}} \sum_{p} \hbar \omega_E f_{BE}^o (\omega_E, T). \tag{3.22}$$

Because the frequency does not depend on wavevector in the Einstein approximation, the entire argument can be brought out of the summation. Then after differentiating with respect to temperature and dividing by volume (i.e., $L^d$), the Einstein specific heat becomes:

$$c_{v,E} = \frac{\hbar\omega_E}{L^d} \frac{\partial f^o_{BE}(\omega_E, T)}{\partial T} \sum_{\mathbf{k}} \sum_p 1$$

$$= \sum_p \eta_a \hbar\omega_E (f^o_{BE})^2 e^{(\hbar\omega_E)/k_B T} \left( \frac{\hbar\omega_E}{k_B T^2} \right)$$

$$= \sum_p \eta_a k_B \frac{\chi_E^2 e^{\chi_E}}{(e^{\chi_E} - 1)^2}, \tag{3.23}$$

where $\eta_a$ is the number of allowable states (i.e., unit cells) per unit real 'volume', and $\chi_E = (\hbar\omega_E)/(k_B T) = \theta_E/T$.

For many problems, the Einstein temperature will be much higher than the real temperature, such that the optical phonon states are sparsely occupied, or $\chi_E \gg 1$. In this limit, Eq. (3.23) gives $c_{v,E} \approx 0$. In such cases, the Debye analysis for only the acoustic branches suffices to characterize the total phonon specific heat. Conversely, for very high temperatures ($\chi_E \ll 1$), we find from $\lim_{\chi_E \to 0} \frac{\chi_E^2 e^{\chi_E}}{(e^{\chi_E} - 1)^2} = 1$ that:

$$c_{v,E} = \sum_p \eta_a k_B \quad (T \gg \theta_E). \tag{3.24}$$

Thus, in spite of the differences with the Debye model, the Einstein result reduces to the Debye model's high-temperature limit (i.e., the Law of Dulong and Petit, see Eq. (3.20)), assuming three phonon branches (i.e., $p = 3$).

A comparison of the Debye and Einstein models is shown in Fig. 3.3, in which both the low-temperature differences and high-temperature convergence are clearly apparent. However, we emphasize that the high-temperature limit for the Einstein model typically requires unusually high temperatures because optical phonon frequencies of most materials correspond to much higher energies than the equivalent thermal energy, i.e., $\omega_E \gg k_B T/\hbar$. However, each of these models can be accurate for all temperatures in the correct context. The important point to remember is that the Debye model is appropriate for acoustic phonons, while the Einstein model (particularly Eq. (3.23)) should be used for optical phonons.

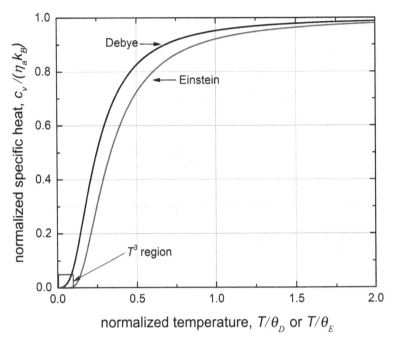

Fig. 3.3    Comparison of specific heat dependence on temperature, as predicted by the Debye and the Einstein model. The specific heats, derived from both the models, converge at low and high temperatures.

### 3.2.3    Electron Specific Heat

The analysis of specific heat for electrons begins with a subtle modification of the internal energy expression of Eq. (3.2):

$$U_e = 2 \sum_{\mathbf{k}} \sum_{p} E_{i,p}(\mathbf{k}) f_{FD}^o \left[ E_{i,p}(\mathbf{k}), T \right], \qquad (3.25)$$

where the pre-factor '2' accounts for spin degeneracy, and the summation over $p$ relates to electronic bands instead of phonon branches. Once again, mathematical convenience dictates the replacement of the $\mathbf{k}$-space summation with an integral by invoking the electron density of states (Eq. (2.22)) and expressing the energy on a volumetric basis:

$$u_e = \int_0^\infty E f_{FD}^o(E, T) D(E) dE, \qquad (3.26)$$

where the factor '2' has now been absorbed into the density of states (cf., Eq. (2.22)).

Another subtlety of the analysis for electrons derives from the Pauli exclusion principle, which dictates that some energy states must be occupied even at absolute zero temperature. The highest such occupied energy is the Fermi energy, $E_F$, which can be defined in terms of the integral form of electron density $\eta_e$:

$$\eta_e = \int_0^\infty f_{FD}^o(E,T)D(E)dE = \int_0^{E_F} D(E)dE, \qquad (3.27)$$

where the latter equality derives from the fact that the Fermi-Dirac function is unity below the Fermi energy and zero above it at zero absolute temperature. In order to keep the derivation more general, we will refrain from deriving an explicit expression for $E_F$, because doing so would require us to assume a specific form of $D(E)$.

To assist in deriving the electron specific heat, an alternative specific internal energy $u*$ can be defined to simplify the subsequent integral analysis:

$$u* \equiv u_e - E_F \eta_e \qquad (3.28)$$

$$= \int_0^\infty E f_{FD}^o(E,T)D(E)dE - \int_0^\infty E_F f_{FD}^o(E,T)D(E)dE$$

$$= \int_0^\infty (E - E_F)f_{FD}^o(E,T)D(E)dE. \qquad (3.29)$$

This contrivance is useful because the subtracted term in Eq. (3.28) is a constant, resulting in a null temperature derivative. Using Eq. (3.27) this constant term can be expressed as an integral involving the distribution function and density of states. The final equality (Eq. (3.29)) contains the difference $(E - E_F)$, which also appears in the distribution function $f_{FD}^o(E,T)$ through the common and broadly valid assumption of equality between the electrochemical potential $\mu$ (which exhibits a significant temperature dependence only at extremely high temperatures for which the thermal energy approaches the Fermi energy; see Eq. (2.11)) and the Fermi energy $E_F$ (which is by definition a constant) (Kittel, 2007).

The electron specific heat can now be expressed by the temperature derivative of $u*$:

$$c_{v,e} = \frac{\partial u}{\partial T} = \frac{\partial u*}{\partial T} = \int_0^\infty (E - E_F)\frac{\partial f_{FD}^o}{\partial T}D(E)dE. \qquad (3.30)$$

The temperature derivative of the Fermi-Dirac distribution function is nonnegligible only in a small region about the Fermi energy. Therefore, the density of states term can be replaced with an integral pre-factor of $D(E_F)$, resulting in:

$$c_{v,e} \approx D(E_F) \int_0^\infty (E - E_F) \frac{\partial f_{FD}^o}{\partial T} dE = k_B^2 T D(E_F) \int_{\frac{-E_F}{k_B T}}^\infty \frac{x^2 e^x}{(e^x + 1)^2} dx,$$

(3.31)

where $x = (E - E_F)/(k_B T)$. Finally, recognizing that typically $E_F \gg k_B T$, the lower bound of the integral can be approximated as $-\infty$, enabling analytical evaluation:

$$c_{v,e} \approx \frac{\pi^2}{3} k_B^2 T D(E_F).$$

(3.32)

Substituting the density of states for three-dimensional free-electron metals (see Eq. (2.22)), the specific heat becomes:

$$c_{v,e} \approx \frac{m k_B^2 \sqrt{2mE_F}}{3\hbar^3} T$$

$$= \frac{\pi^2 k_B^2 \eta_e}{2E_F} T,$$

(3.33)

where the latter equality derives from the definition of the Fermi energy for a parabolic band (Eq. (1.40)). A distinguishing feature of the final result is the linear temperature dependence, which can be used to assess the relative contributions of electrons ($\sim T^1$) and phonons ($\sim T^3$) in metals at low temperatures (i.e., well below the Debye and Fermi temperatures).

### 3.2.4 *Specific Heat for Low-Dimensional Structures*

The specific heat integral in **k**-space (Eq. (3.5)) can be converted to frequency space generally (i.e., without invoking a dispersion assumption such as the Debye approximation) for phonons through the use of the density of states:

$$c_v = \frac{\partial u}{\partial T} = \sum_p \int \frac{E_{i,p}(\mathbf{k})}{(2\pi)^d} \frac{\partial f_{BE}^o}{\partial T} d\mathbf{k}$$

$$= \sum_p \int \hbar\omega D_{dD}(\omega) \frac{\partial f_{BE}^o}{\partial T} d\omega,$$

(3.34)

where $d$ represents the dimensionality of the problem (i.e., "$dD = 1D$" for $d = 1$).

Then, applying the Debye model's density of states, a general expression for the Debye specific heat of acoustic phonons becomes:

$$c_{v,D} = d \times \eta_a k_B \sum_p \left( \frac{T}{\theta_{D,p}} \right)^d \int_0^{\theta_{D/T}} \frac{x^{d+1}e^x dx}{(e^x - 1)^2}. \tag{3.35}$$

This result reinforces the memorable result that the specific heat for low temperatures is proportional to temperature raised to the power of the dimensionality for temperatures well below the Debye temperature, as shown in Fig. 3.4. The results indicate that the power law relationship $c_v \sim T^d$ holds well up to $T \approx 0.1\theta_D$. Consequently, the temperature dependence of specific heat (a property that is relatively easy to measure) provides a means of assessing the effective dimensionality of the medium under study.

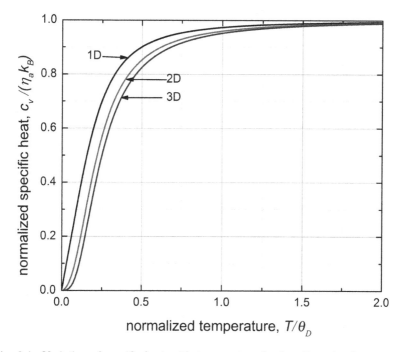

**normalized temperature,** $T/\theta_D$

Fig. 3.4 Variation of specific heat with temperature for low-dimensional structures, as predicted by the Debye model. Notice the $T^d$ dependence of specific heat at low temperatures, where $d$ is the dimensionality of the medium.

## 3.3    Thermal Conductivity from Kinetic Theory

The foregoing expressions for specific heat serve important roles in determining a second prominent thermal property—thermal conductivity. Various approaches are available to derive the latter quantity, and here we offer the most common and intuitive derivation. The subsequent chapters consider other derivation approaches that are generally more rigorous and versatile.

Kinetic theory covers broadly the behavior of particles in an ensemble and can be used to derive many thermophysical properties (Vincenti and Kruger, 1967). The basic theory treats particles as independent entities that can collide, or scatter, with each other or with other objects such as defects and boundaries. As such, the approach is somewhat agnostic to the type of particle, as long as its velocity and ability to carry a property (such as thermal energy) are known.

Figure 3.5 shows this scenario schematically. An energy-carrying particle (e.g., electron or phonon in the present context) sits at the vertical position $z + \Lambda_z$ within a field of particles with average intensive internal energy $u$ that depends on position $z$. The particles move in three dimensions randomly and experience a collision one time for every distance $\Lambda$ traveled, on average. The distance $\Lambda$ is termed the *mean free path*, or scattering length. In the case of the particle highlighted in Fig. 3.5, the vertical ($z$) component of its distance traveled before its next collision is $\Lambda_z$, as its direction makes a polar angle $\theta$ with the $z$ axis.

The heat flux rate (per unit area) can be expressed in terms of the $z$-components of the particle velocity and mean free path:

$$q_z'' = \frac{1}{2}v_z \left[ u \left( z - \Lambda_z \right) - u \left( z + \Lambda_z \right) \right],  \tag{3.36}$$

where the $\frac{1}{2}$ term derives from the fact that only half the particles move up from $z - \Lambda_z$ or down from $z + \Lambda_z$, and $v_z$ is the $z$-component of the particle's velocity. The energy difference in Eq. (3.36) can be expanded as a Taylor series:

$$u(z + \Lambda_z) = u(z - \Lambda_z) + \left. \frac{\partial u}{\partial z} \right|_z (2\Lambda_z) + \vartheta \left( \Lambda_z^2 \right).  \tag{3.37}$$

Using $\Lambda_z = \Lambda \cos \theta$ and $v_z = v \cos \theta$, the heat flux becomes:

$$q_z'' \approx -v_z \Lambda_z \frac{\partial u}{\partial z} = - \left( \cos^2 \theta \right) v \Lambda \frac{\partial u}{\partial z}.  \tag{3.38}$$

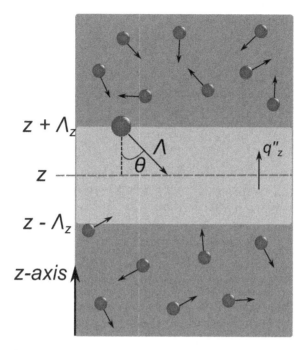

Fig. 3.5   A schematic depicting the kinetic theory of thermal conductivity. An atom at $z + \Lambda_z$ travels a distance equivalent to its mean free path, $\Lambda$ ($\Lambda_z$ in the z-direction), before experiencing a collision. This atomic motion results in a heat flux, along the z-direction, which is a function of the particle velocity and the particle mean free path.

The foregoing steps assumed a specific direction of motion, but the actual directions within the ensemble are randomized. Therefore, an average heat flux must be defined by integrating over all possible directions through the three-dimensional solid angle $d\Omega = \sin\theta d\theta d\psi$, where $\psi$ is the azimuthal angle:

$$\langle q''_z \rangle = -v\Lambda \frac{\partial u}{\partial z} \left[ \frac{1}{2\pi} \int_0^{2\pi} \int_0^{\frac{\pi}{2}} \cos^2\theta \sin\theta d\theta d\psi \right]$$

$$= -\frac{1}{3} v\Lambda \frac{\partial u}{\partial z}. \tag{3.39}$$

Assuming that the scattering processes are frequent enough to establish local thermodynamic equilibrium (which is *not* the case for predominantly ballistic transport), the chain rule can be applied to convert the energy gradient to a temperature gradient:

$$\frac{\partial u}{\partial z} = \frac{\partial u}{\partial T}\frac{\partial T}{\partial z}. \tag{3.40}$$

Importantly, the first term on the right side is the previously developed specific heat (Eq. 3.1), and the average heat flux can be expressed as:

$$\langle q_z'' \rangle = -\frac{1}{3}v\Lambda\frac{\partial u}{\partial T}\frac{\partial T}{\partial z} = -\underbrace{\frac{1}{3}c_v v\Lambda}_{\kappa}\frac{\partial T}{\partial z}, \tag{3.41}$$

where the final form matches that of the classical Fourier's law, $q_z'' = -\kappa(\partial T/\partial z)$. The foregoing derivation therefore relates a material's thermal conductivity $\kappa$ to the specific heat, velocity, and mean free path of thermal energy carriers:

$$\kappa = \frac{1}{3}c_v v\Lambda. \tag{3.42}$$

Some important issues and caveats concerning this expression follow:

- Thermal conductivity inherits the temperature dependence of the specific heat, velocity, and mean free path. We have considered elements of the first two, the last remains for the subsequent two chapters.

- The derivation above was somewhat casual regarding the variability of carrier velocity, which depends on the distribution function and occupation statistics. We will consider these issues further in the next chapter.

- For very small materials, any or all of the three components can be influenced substantially by the size of the domain under study, its lattice and defect structure, and its temperature.

## Example Problems

**Problem 3.1: Graphene ZA branch specific heat**

In this chapter, we obtained integral expressions for the specific heat of branches that can be approximated with a linear dispersion (Debye model) and constant dispersion (Einstein model). The ZA branch of graphene, which represents out-of-plane vibrations (see Appendix), is however closely approximated near the Brillouin zone center by a quadratic dispersion relation of the form $\omega = CK^2$ where $C$ is a constant.

(a) Determine the maximum cutoff wavevector $K_Q$ and the corresponding cutoff frequency $\omega_Q$ in terms of the unit cell density $\eta_a$.

(b) Obtain an integral expression for the specific heat of the ZA branch as a function of temperature.

The low temperature specific heat of graphene shows a linear dependence on temperature (see Fig. 3.6) which then becomes quadratic for temperatures greater than 100 K. Can you explain this behavior based on your knowledge of the dispersion relation of graphene (see Chapter 2 Examples for a plot) and the expression you have just obtained in this problem?

*Solution*

(a) The cutoff wavevector $K_Q$ is found by equating the number of states in **k**-space within a circle of radius $K_Q$ to the total number of unit cells $N$.

$$\frac{\pi K_Q^2}{\left(\frac{2\pi}{L}\right)^2} = N, \quad K_Q = (4\pi\eta_a)^{1/2}, \tag{3.43}$$

where $\eta_a$ is the number of unit cells per unit area. The cutoff frequency $\omega_Q$ is then obtained from the dispersion relation:

$$\omega_Q = 4C\pi\eta_a. \tag{3.44}$$

(b) The specific heat $c_{v,ZA}$ is given by:

$$c_{v,ZA} = \int_0^{\omega_Q} \hbar\omega \frac{\partial f_{BE}^o}{\partial T} D_{Q,2D}(\omega)d\omega, \tag{3.45}$$

where $D_{Q,2D}(\omega)$ is the two-dimensional density of states under the quadratic dispersion model:

$$D_{Q,2D}(\omega) = \frac{1}{L^2}\frac{dN}{d\omega}$$

$$= \frac{1}{L^2}\frac{dN}{dK}\frac{dK}{d\omega}$$

$$= \frac{1}{L^2}\frac{d}{dK}\left(\frac{\pi K^2}{(2\pi/L)^2}\right)\frac{1}{2CK}$$

$$= \frac{1}{4\pi C}. \tag{3.46}$$

Thus the DOS of the ZA branch is a constant. Substituting the above expression into the integral in Eq. (3.45), and using the derivative of the Bose-Einstein distribution function, we obtain:

$$c_{v,ZA} = \frac{1}{4\pi C}\int_0^{\omega_Q}\hbar\omega\frac{\exp(\hbar\omega/k_BT)}{(\exp(\hbar\omega/k_BT)-1)^2}\frac{\hbar\omega}{k_BT^2}d\omega$$

$$= \eta_a k_B\left(\frac{T}{\theta_Q}\right)\int_0^{\theta_Q/T}\frac{x^2 e^x}{(e^x-1)^2}dx, \tag{3.47}$$

where $x = \hbar\omega/k_BT$ and $\theta_Q = \hbar\omega_Q/k_B$ (analogous to the Debye temperature $\theta_D$). For temperatures much less than $\theta_Q$, the upper limit of the integral in Eq. (3.47) can be taken to be $\infty$, and the specific heat is proportional to $T$. This result explains the linear dependence of the specific heat of graphene at low temperatures. At higher temperatures, the ZA branch becomes fully populated and the linear LA and TA modes contribute to specific heat. This explains the quadratic dependence at higher temperatures (see Section 3.2.4).

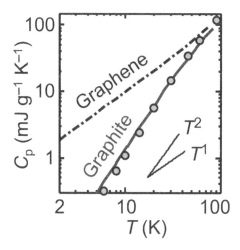

Fig. 3.6   Temperature dependence of the specific heat of graphene and graphite. Figure originally published by Pop *et al.* (2012). Used with permission.

---

### Problem 3.2: Specific heat of metals

Figure 3.7 shows experimental measurements of the heat capacity of potassium at low temperatures. The following temperature dependence is observed:

$$c_v/T = 2.08 + 2.57T^2,$$

where $c_v/T$ has units of mJ/mol K$^2$ and $T$ is in K.

(a) Provide analytical expressions for the y-intercept and slope of the graph. Hint: Neglect any optical phonon contribution to specific heat as the experimental data are provided for low temperatures.

(b) Assuming that the conduction electron density in potassium is $1.34 \times 10^{22}$ cm$^{-3}$, determine the Fermi energy of potassium. Note that the experimental data are expressed per mole of potassium, while the heat capacity expressions derived in this chapter are per unit volume. Assume the density and atomic mass of potassium are 0.862 g/cc and 39 amu respectively.

(c) Potassium has a body-centered cubic (BCC) structure (1 atom per primitive unit cell) with an atomic density of $1.33 \times 10^{22}$

atoms/cm$^3$. Determine the Debye temperature of potassium assuming that the three acoustic branches are replaced by a single branch of uniform group velocity.

*Solution*

(a) Specific heat $c_v$ can be expressed as a sum of electron and phonon contributions:

$$c_v = c_{v,e} + c_{v,p}$$

$$= \frac{\pi^2 k_B^2 \eta_e}{2E_F} T + \frac{234 \eta_a k_B}{\theta_D^3} T^3. \tag{3.48}$$

Hence,

$$\frac{c_v}{T} = \underbrace{\frac{\pi^2 k_B^2 \eta_e}{2E_F}}_{y-\text{intercept}} + \underbrace{\frac{234 \eta_a k_B}{\theta_D^3}}_{\text{slope}} T^2. \tag{3.49}$$

(b) Using the experimental data and the result from part (a),

$$2.08 \text{ mJ/mol K}^2 = 4.59 \times 10^{-5} \text{ J/cc K}^2 = \frac{\pi^2 k_B^2 \eta_e}{2E_F}. \tag{3.50}$$

Substituting $k_B = 1.3806 \times 10^{-23}$ J/K, $\eta_e = 1.34 \times 10^{22}$ cm$^{-3}$, we obtain $E_F = 2.74 \times 10^{-19}$ J $= 1.71$ eV.

(c) From the slope of the given graph and the result in part (a),

$$2.57 \text{ mJ/mol K}^4 = 5.68 \times 10^{-5} \text{ J/cc K}^4 = \frac{234 \eta_a k_B}{\theta_D^3}. \tag{3.51}$$

Substituting $\eta_a = 1.33 \times 10^{22}$ atoms/cm$^3$, we obtain $\theta_D = 91.1$ K.

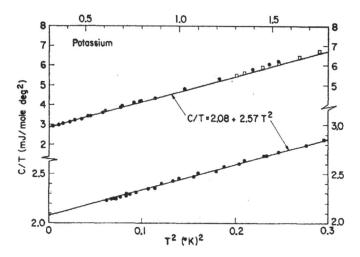

Fig. 3.7 Temperature dependence of the specific heat of potassium and sodium. Figure originally published by [Lien and Phillips (1964)]. Used with permission.

## Problem 3.3: Thermal conductivity from kinetic theory

In this chapter, we derived the thermal conductivity of a three-dimensional material from kinetic theory. Perform a similar analysis for one- and two-dimensional materials to obtain the following generalized expression:

$$\kappa = \frac{1}{d} c_v v \Lambda,$$

where $d$ is the dimension and can take the values 1, 2 or 3. Also derive an integral expression for the thermal conductivity and observe the temperature dependence at low temperatures. Assume that the velocity and mean free path are independent of temperature and carrier energy.

### Solution

The following expression for heat flux $q_z''$ was obtained in the chapter:

$$q_z'' \approx -v_z \Lambda_z \frac{\partial u}{\partial z} = -v_z \Lambda_z \frac{\partial u}{\partial T} \frac{\partial T}{\partial z} = -c_v v_z \Lambda_z \frac{\partial T}{\partial z}. \tag{3.52}$$

In 1D, $v_z = v$, $\Lambda_z = \Lambda$.

$$q_z'' = -c_v v \Lambda \frac{\partial T}{\partial z}. \tag{3.53}$$

In 2D, $v_z = v \cos\theta$, $\Lambda_z = \Lambda \cos\theta$. We average the heat flux over an angle of $\pi$ radians.

$$q_z'' = -c_v v \Lambda \frac{\partial T}{\partial z} \frac{1}{\pi} \int_0^\pi \cos^2\theta d\theta$$

$$= -\frac{1}{2} c_v v \Lambda \frac{\partial T}{\partial z}. \tag{3.54}$$

In 3D, the heat flux is averaged over a solid angle of $2\pi$ steradians.

$$q_z'' = -c_v v \Lambda \frac{\partial T}{\partial z} \frac{1}{2\pi} \int_0^{2\pi} \int_0^{\pi/2} \cos^2\theta \sin\theta d\theta d\psi$$

$$= -\frac{1}{3} c_v v \Lambda \frac{\partial T}{\partial z}. \tag{3.55}$$

From Eqs. (3.53)–(3.55), thermal conductivity $\kappa$ in $d$ dimensions is given by,

$$\kappa = \frac{1}{d} c_v v \Lambda. \tag{3.56}$$

Substituting the low temperature result for specific heat $c_v$, we find:

$$\kappa = \eta_a k_B v \Lambda \left( \frac{T}{\theta_D} \right)^d \int_0^\infty \frac{x^{d+1} e^x dx}{(e^x - 1)^2}. \tag{3.57}$$

Note that the Debye approximation is used in the above expression for specific heat. Also we have neglected multiple phonon polarizations. At low temperatures, the thermal conductivity shows the same temperature dependence as the specific heat and scales as $T^d$.

**Problem 3.4: Specific heat of a diatomic chain**

Consider the diatomic chain studied in Chapter 1 with atomic masses $m_1$, $m_2$ $(m_2 > m_1)$ and a uniform atom spacing of $a$. Also assume a uniform spring constant $g$ between all adjacent atoms. In this problem, we calculate the specific heat of the diatomic chain using the Debye model for the acoustic branch and the Einstein model for the optical branch. Assume that the constant frequency $\omega_E$ in the Einstein model is an average of the minimum and maximum frequencies of the optical branch.

(a) Show that the ratio of Einstein and Debye temperatures can be expressed in terms of the mass ratio $m_2/m_1$ as follows:

$$\frac{\theta_E}{\theta_D} = \frac{1}{\pi} \left( \sqrt{\frac{m_1}{m_2}} + \sqrt{\frac{m_2}{m_1}} + \sqrt{1 + \frac{m_2}{m_1}} \right).$$

(b) Calculate the normalized acoustic and optical phonon specific heats (normalized by $\eta_a k_B$) at normalized temperatures of $T/\theta_D = 0.2, 1$ and 2. Assume a mass ratio $m_2/m_1 = 2$. Also provide an intuitive explanation of your numerical results.

(c) Use the online Chapter 3 CDF tool[1] to evaluate the acoustic and optical contributions to the total specific heat as a function of temperature. Also observe how these contributions change with varying mass ratio. Again, provide a physical explanation for the trend in the curves with varying mass ratio.

*Solution*

(a) The Einstein frequency $\omega_E$ is calculated by taking an average of the minimum and maximum frequencies of the optical branch. See Section 1.6 for derivations of the minimum $(\omega_+(K = \pi/a))$ and maximum $(\omega_+(K = 0))$ frequencies.

$$\omega_E = \frac{1}{2}(\omega_+(K = 0) + \omega_+(K = \pi/a))$$

$$= \frac{1}{2} \left( \sqrt{\frac{2g}{\mu}} + \sqrt{\frac{2g}{m_1}} \right)$$

$$= \sqrt{\frac{g}{2}} \left( \frac{\sqrt{m_1 + m_2} + \sqrt{m_2}}{\sqrt{m_1 m_2}} \right). \tag{3.58}$$

---

[1] See http://nanohub.org/groups/cdf_tools_thermal_energy_course/wiki

The Debye frequency $\omega_D$ is just the product of the group velocity of the acoustic branch at the Brillouin zone center and the Debye cutoff wavevector $K_D$. In 1D, $K_D = \pi\eta_a = \pi/a$ since the unit cell density is $1/a$. See Section 1.6 for a derivation of the group velocity of the acoustic branch at the center of Brillouin zone.

$$\omega_D = v_g(K = 0)K_D$$

$$= a\sqrt{\frac{g\mu}{2m_1 m_2}}\frac{\pi}{a}$$

$$= \pi\sqrt{\frac{g}{2(m_1 + m_2)}}. \tag{3.59}$$

From Eqs. (3.58) and (3.59), the ratio of Einstein and Debye temperatures is given by:

$$\frac{\theta_E}{\theta_D} = \frac{\omega_E}{\omega_D} = \frac{1}{\pi}\left(\sqrt{\frac{m_1}{m_2}} + \sqrt{\frac{m_2}{m_1}} + \sqrt{1 + \frac{m_2}{m_1}}\right). \tag{3.60}$$

(b) The specific heat of the acoustic branch is given by (see Section 3.2.4):

$$c_{v,D} = \eta_a k_B \left(\frac{T}{\theta_D}\right)\int_0^{\theta_D/T} \frac{x^2 e^x dx}{(e^x - 1)^2}. \tag{3.61}$$

For $T/\theta_D = 0.1$, 1 and 2, $c_{v,D}/\eta_a k_B = 0.328$, 0.973 and 0.993 respectively (the integral was evaluated numerically). Observe that the specific heat is very close to the Dulong and Petit limit of $c_{v,D} = \eta_a k_B$ for temperatures higher than the Debye temperature. The specific heat of the optical branch is given by (see Section 3.2.2):

$$c_{v,E} = \eta_a k_B \frac{\chi_E^2 e^{\chi_E}}{(e^{\chi_E} - 1)^2}, \tag{3.62}$$

where $\chi_E = \theta_E/T$. For $m_2/m_1 = 2$, $\theta_E/\theta_D = 1.226$ (using the result derived in part (a) of this problem). Hence $\theta_E/T = 1.226\theta_D/T$. For $T/\theta_D = 0.1$, 1 and 2, $\chi_E = 12.26$, 1.226 and 0.613 respectively. Thus $c_{v,E}/\eta_a k_B = 0.0007$, 0.884 and 0.969 for $T/\theta_D = 0.1$, 1 and 2 respectively. Note that the optical phonon

Table 3.1 Acoustic and optical phonon specific heats of a diatomic chain with $m_2/m_1 = 2$.

| $\dfrac{T}{\theta_D}$ | $\dfrac{c_{v,D}}{\eta_a k_B}$ | $\dfrac{c_{v,E}}{\eta_a k_B}$ | $\dfrac{c_{v,D}}{c_{v,D} + c_{v,E}} \times 100\%$ |
|---|---|---|---|
| 0.1 | 0.328 | 0.0007 | 99.78% |
| 1 | 0.973 | 0.884 | 52.4% |
| 2 | 0.993 | 0.969 | 50.61% |

specific heat is almost zero for $T/\theta_D = 0.1$. This is because the high frequency optical mode is negligibly populated at such low temperatures. The optical phonon specific heat also approaches the Dulong and Petit law for high temperatures.

Table 3.1 shows a summary of the results calculated. At low temperatures, the heat capacity of the acoustic branch dominates that of the optical branch. For temperatures above the Debye temperature, $c_{v,D}/c_{v,E} \approx 1$ indicating that both the acoustic and optical modes contribute equally to specific heat.

(c) Figure 3.8 shows snapshots from the online Chapter 3 CDF tool where the acoustic and optical contributions to the total specific heat are plotted as a function of temperature. Clearly the acoustic branch dominates the specific heat at low temperatures, and the fractional contributions tend to 0.5 for very high temperatures. As the mass ratio increases, the Einstein frequency moves farther from the Debye frequency. Hence the temperature at which the acoustic and optical contributions become equal also increases with increasing mass ratio.

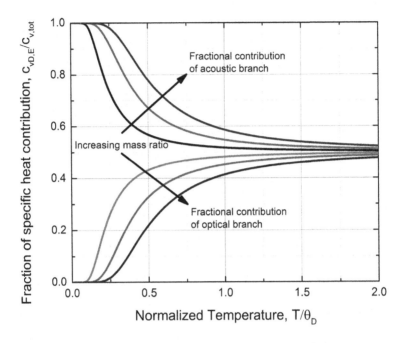

Fig. 3.8    Acoustic and optical contributions to specific heat.

Chapter 4

# Landauer Transport Formalism

This chapter derives fundamental limits of heat transfer carried by phonons and electrons between two contacts, or reservoirs. A general Landauer formalism is employed in combination with an enumeration of available electronic and vibrational states for a given problem. A significant portion of the chapter focuses on spectral conductance—the thermal conductance per unit spectral quantity such as energy, frequency, wavelength, or wavenumber. The understanding of this spectral behavior ultimately enables the engineering of a device through material size by, for example, suppressing conduction at a given wavelength.

## 4.1 Basic Theory

We begin by establishing how thermal energy can be stored in a reservoir and the relevant wavelengths of energy carriers in those reservoirs. Both phonons and electrons store energy through a distribution of energy states, but these distributions differ because of the restrictions (or lack thereof) on the number carriers that can occupy each state, as described in Chapter 2. Electrons are governed by the Fermi-Dirac distribution function, which restricts each state to hold up to two electrons, whereas the number of phonons in a given state is unlimited, resulting in the Bose-Einstein distribution.

Our interest is to develop relations for thermal transport between two thermal reservoirs connected by a device, as shown in Fig. 4.1(a) for the general problem. To illustrate the calculation of net heat flux, we first address transport in the simplified all-1D structure of Fig. 4.1(b).

The total heat flux (i.e., heat flow rate per unit contact 'area', with units of $W/m^{d-1}$, where $d$ is the problem dimension) from the left reservoir

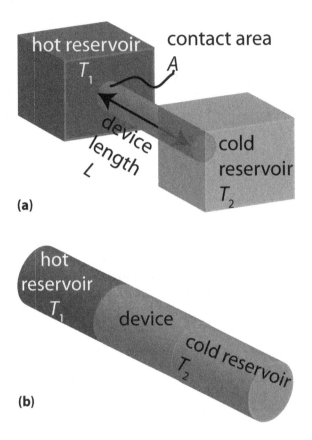

Fig. 4.1   Schematic of a general contact-device-contact arrangement with (a) 3D (bulk) and (b) 1D (wire) contacts.

to the right can be calculated by summing the product of energy density and velocity over all wavevectors with positive $x$-components as

$$J_{Q,L \to R}(T_1) = \frac{1}{L^d} \sum_p \sum_{\mathbf{k};k_x>0} v_{gx,p}(\mathbf{k}) \mathcal{T}_p(\mathbf{k})$$

$$\times [E_{i,p}(\mathbf{k}) - \mu] [f_i^o(E_{i,p}(\mathbf{k}), T_1) + c_0], \qquad (4.1)$$

where $\mathcal{T}_p(\mathbf{k})$ represents the transmission function, which accounts for the probability of transport through the device,[1] and the $x$ direction lies along

---

[1] We note that the companion books in this series by Datta (2012) and Lundstrom and Jeong (2013) use the Latin symbol $T$ for transmission function; here we use the modified $\mathcal{T}$, because $T$ holds a reserved place for thermal scientists and engineers.

the axis of the device. For 3D systems, $J_Q$ is equivalent to the heat flux $q''$ of the previous chapter. Note that the presence of the chemical potential $\mu$ accounts for the equilibrium energy levels in the reservoir and the associated redistribution that occurs to maintain thermodynamic equilibrium when a carrier leaves the reservoir.

A reciprocal expression to Eq. (4.1) can be formed for heat flow from the right reservoir to the left, and the net heat flux can then be expressed as their sum:

$$J_{Q,\text{net}} = \frac{1}{L^d} \sum_p \sum_{\mathbf{k};k_x>0} v_{gx,p} \mathcal{T}_p \left[E_{i,p} - \mu\right] \left[f_i^o(T_1) + c_0\right]$$

$$+ \frac{1}{L^d} \sum_p \sum_{\mathbf{k};k_x<0} v_{gx,p} \mathcal{T}_p \left[E_{i,p} - \mu\right] \left[f_i^o(T_2) + c_0\right]$$

$$= \frac{1}{L^d} \sum_p \sum_{\mathbf{k};k_x>0} v_{gx,p} \mathcal{T}_p \left[E_{i,p} - \mu\right] \left[f_i^o(T_1) - f_i^o(T_2)\right], \qquad (4.2)$$

where the energy and wavevector dependencies of the various terms are now and hereafter implied, and the group velocity is negative for $k_x < 0$. Notably, we see that the $c_0$ terms related to the zero-point energy (see Eq. (2.12)) cancel each other. This effect suggests that the assessment of the 'dominant' phonon wavelength for thermal transport discussed later should exclude this term, as its effect is nullified by carriers moving in opposing directions.

Our interest is to study transport processes, and as such, the double summation in Eq. (4.2) is cumbersome. In particular, we would like to turn the summation over **k**-space (i.e., over all active wavelengths and directions) into an integral. The summation over polarization branches (e.g., longitudinal-acoustic, transverse-optical, etc. for phonons) normally must remain (or else be subject to approximation), but the summation over **k**-space can be converted to a Landauer integral according to the dimensionality of the problem:

$$J_{Q,\text{net}} = \frac{1}{L^d} \sum_p \sum_{\mathbf{k};k_x>0} v_{gx,p} \mathcal{T}_p \left[E_{i,p}(k) - \mu\right] \left[f_i^o(T_1) - f_i^o(T_2)\right], \qquad (4.3)$$

$$(1\text{D}) = \sum_p \int_0^\infty \frac{v_{g,p} \mathcal{T}_p \left[E_{i,p}(k) - \mu\right]}{2\pi} \left[f_i^o(T_1) - f_i^o(T_2)\right] dk, \qquad (4.4)$$

$$(2D) = \sum_p \int_{-\frac{\pi}{2}}^{\frac{\pi}{2}} \int_0^\infty \frac{v_{g,p} \cos\theta \mathcal{T}_p \left[E_{i,p} - \mu\right]}{4\pi^2} \left[f_i^o(T_1) - f_i^o(T_2)\right] k\,dk\,d\theta$$

$$= \sum_p \int_0^\infty \frac{v_{g,p} \mathcal{T}_p \left[E_{i,p}(k) - \mu\right]}{2\pi^2} \left[f_i^o(T_1) - f_i^o(T_2)\right] k\,dk, \qquad (4.5)$$

$$(3D) = \sum_p \int_0^{2\pi} \int_0^{\frac{\pi}{2}} \int_0^\infty \frac{v_{g,p} \cos\theta \mathcal{T}_p \left[E_{i,p} - \mu\right]}{8\pi^3}$$

$$\times \left[f_i^o(T_1) - f_i^o(T_2)\right] k^2 dk \sin\theta\, d\theta\, d\psi$$

$$= \sum_p \int_0^\infty \frac{v_{g,p} \mathcal{T}_p \left[E_{i,p}(k) - \mu\right]}{8\pi^2} \left[f_i^o(T_1) - f_i^o(T_2)\right] k^2 dk, \qquad (4.6)$$

where $\theta$ is the angle formed by the $x$-axis and the velocity ($\overrightarrow{k}$) direction, and Eqs. (4.5) and (4.6) assume that transmission function is independent of direction. Notably, the explicit domain 'volumes' $L^d$ in Eq. (4.2) cancel in the conversion to an integral. Also of importance for later use is the derivative of the distribution function with respect to temperature, which was provided in Eq. (3.6). This function will be used extensively in formulations of thermal conductance.

The heat flux is commonly expressed in terms of an integral over frequency (phonons) or energy (electrons) instead of the generic **k**-space used above (which applies to either phonons or electrons). For phonons, the conversion to frequency space involves the introduction of the density of states per unit frequency $D(\omega)$ (see Eqs. (2.19)–(2.21)). With the null chemical potential $\mu$ associated with bosons and substituting $\hbar\omega = E$, the dimension-specific heat fluxes become:

$$J_{Q,\text{ph}} =$$

$$(1D) = \sum_p \int_0^\infty \frac{1}{2} v_{g,p} D_{1D}(\omega) \mathcal{T}_p(\omega) \hbar\omega \left[f_{BE}^o(T_1) - f_{BE}^o(T_2)\right] d\omega, \qquad (4.7)$$

$$(2D) = \sum_p \int_0^\infty \frac{1}{2} \frac{2 v_{g,p}}{\pi} D_{2D}(\omega) \mathcal{T}_p(\omega) \hbar\omega \left[f_{BE}^o(T_1) - f_{BE}^o(T_2)\right] d\omega, \quad (4.8)$$

$$(3D) = \sum_p \int_0^\infty \frac{1}{2} \frac{v_{g,p}}{2} D_{3D}(\omega) \mathcal{T}_p(\omega) \hbar\omega \left[f_{BE}^o(T_1) - f_{BE}^o(T_2)\right] d\omega, \qquad (4.9)$$

where we have now made explicit the possible dependence of the transmission function on frequency.

For electrons, we retain the finite chemical potential $\mu$, use the density of states (Eqs. (2.23)–(2.25)), and integrate over energy rather than frequency:

$$J_{Q,\text{el}} =$$

$$(1D) = \int_0^\infty \frac{1}{2} v_g D_{1D}(E) \mathcal{T}(E)(E - \mu) \left[ f_{FD}^o(T_1) - f_{FD}^o(T_2) \right] dE, \quad (4.10)$$

$$(2D) = \int_0^\infty \frac{1}{2} \frac{2v_g}{\pi} D_{2D}(E) \mathcal{T}(E)(E - \mu) \left[ f_{FD}^o(T_1) - f_{FD}^o(T_2) \right] dE, \quad (4.11)$$

$$(3D) = \int_0^\infty \frac{1}{2} \frac{v_g}{2} D_{3D}(E) \mathcal{T}(E)(E - \mu) \left[ f_{FD}^o(T_1) - f_{FD}^o(T_2) \right] dE, \quad (4.12)$$

where we have eliminated the summation, which is unnecessary for electrons with non-overlapping bands.

## 4.2 Number of Modes

The grouping of terms in the heat flux relations of Eq. (4.7)–(4.12) provides a useful basis for understanding the underlying physics. Lundstrom and Jeong (2013) provide an elegant parallel treatment for electrical current that we will adapt here for thermal current, or $Q$ (in W), which is the product of the thermal current density $J_Q$ and 'area' (null for 1D, contact width $W$ for 2D, and contact area $A$ for 3D). For phonons, a general form for thermal current, or heat flow rate, is:

$$Q_{\text{ph}} = \frac{1}{2\pi} \int_0^\infty M(\omega) \mathcal{T}(\omega) \hbar\omega \left[ f_{BE}^o(T_1) - f_{BE}^o(T_2) \right] d\omega, \quad (4.13)$$

where $M(\omega)$ is called the 'number of modes' and represents the number of carrier half-wavelengths that fit into the contact 'area'.[2] For example, in purely one-dimensional transport, the contact can fit only one carrier in its cross section, and $M(\omega)$ is simply 1. For two-dimensional problems (e.g.,

---

[2]This result is mathematically identical to that derived by Lundstrom and Jeong (2013) for phonons, who express the variable of integration as $\hbar\omega$.

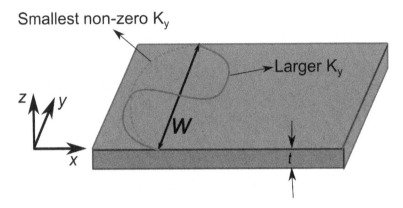

Fig. 4.2   Schematic of the number of modes $M$. This factor is essentially the number of half-waves for a carrier with wavelength $\lambda$ (and corresponding energy $E$ derived from the carrier's dispersion) that fit into a cross-section of the device perpendicular to the direction of transport.

a thin film device), the number of modes will increase in proportion to the contact length $W$, as shown in Fig. 4.2.

The corresponding expressions for each dimensionality are:

$$M(\omega) =$$

$$(\text{1D}) = M_{1D}(\omega), \qquad (4.14)$$

$$(\text{2D}) = W M_{2D}(\omega), \qquad (4.15)$$

$$(\text{3D}) = A M_{3D}(\omega). \qquad (4.16)$$

Then, using $J_Q(\text{1D}) = Q$, $J_Q(\text{2D}) = Q/W$, and $J_Q(\text{3D}) = Q/A$ and comparing to the phonon expressions in Eqs. (4.7)–(4.9), we can deduce the following relations for mode densities:

$$M_{1D}(\omega) = 1 = \pi[v_g(\omega)]D_{1D}(\omega), \qquad (4.17)$$

$$M_{2D}(\omega) = \frac{K(\omega)}{\pi} = \pi\left[\frac{2v_g(\omega)}{\pi}\right]D_{2D}(\omega), \qquad (4.18)$$

$$M_{3D}(\omega) = \frac{K(\omega)^2}{4\pi} = \pi\left[\frac{v_g(\omega)}{2}\right]D_{3D}(\omega), \qquad (4.19)$$

.where we repeat the frequency-based phonon density of states from Chapter 2 (Eqs. (2.19)–(2.21)), given their importance here:

$$D_{1D}(\omega) = \frac{1}{L}\frac{dN_{1D}}{d\omega} = \frac{1}{L}\frac{dN_{1D}}{dK}\frac{dK}{d\omega} = \frac{1}{v_g(\omega)\pi}, \tag{4.20}$$

$$D_{2D}(\omega) = \frac{1}{L^2}\frac{dN_{2D}}{d\omega} = \frac{1}{L^2}\frac{dN_{2D}}{dK}\frac{dK}{d\omega} = \frac{K(\omega)}{2\pi v_g(\omega)}, \tag{4.21}$$

$$D_{3D}(\omega) = \frac{1}{L^3}\frac{dN_{3D}}{d\omega} = \frac{1}{L^3}\frac{dN_{3D}}{dK}\frac{dK}{d\omega} = \frac{K(\omega)^2}{2\pi^2 v_g(\omega)}. \tag{4.22}$$

The combination of Eqs. (4.17)–(4.22) gives a complete expression for the number of phonon modes in each dimensionality:

$$M(\omega) =$$

$$(1D) = \pi[v_g(\omega)]\frac{1}{v_g(\omega)\pi} = 1, \tag{4.23}$$

$$(2D) = W\pi\left[\frac{2v_g(\omega)}{\pi}\right]\frac{K(\omega)}{2\pi v_g(\omega)}, \tag{4.24}$$

$$(3D) = A\pi\left[\frac{v_g(\omega)}{2}\right]\frac{K(\omega)^2}{2\pi^2 v_g(\omega)}. \tag{4.25}$$

The foregoing equations can be altered to demonstrate the interpretation of the number of modes as representing the maximum number of half-wavelengths in the device cross-section. Using $K = 2\pi/\lambda$, Eqs. (4.24) and (4.25) become:

$$M(\omega) =$$

$$(2D) = \frac{W}{\lambda/2}, \tag{4.26}$$

$$(3D) = \frac{A}{\frac{4}{\pi}\left(\frac{\lambda}{2}\right)^2}, \tag{4.27}$$

where the factor $4/\pi$ in the latter equation is related to the number of wavelengths that fit into a circle of area $A$ (whose area is $\pi/4\times$ diameter$^2$).

The mode number results also reveal an unexpected outcome, namely that the terms in square brackets of Eqs. (4.24) and (4.25) represent the space-averaged $x$-component of group velocity for a given frequency $\omega$. In 1D, the average is the same as the group velocity because motion is constrained to a single direction. In 2D, the average of $\cos\theta$ around a semicircle in the positive $x$ direction is $2/\pi$. In 3D, the corresponding average about the solid angle hemisphere of $2\pi$ steradians is $1/2$.

Similar analysis for electrons produces a result for electronic heat flow, which in the following is expressed in the customary manner as an integral in energy space:

$$Q_{\text{el}} = \frac{1}{\pi\hbar} \int_0^\infty M(E)\mathcal{T}(E)\,(E - \mu)\,[f_{FD}^o(T_1) - f_{FD}^o(T_2)]\,dE, \qquad (4.28)$$

where the number of modes for electrons in a parabolic conduction band has been derived by Lundstrom and Jeong (2013, Eq. (2.31)):

$$M(E) =$$

$$(1\text{D}) = M_{1D}(E)$$

$$= H(E - E_c), \qquad (4.29)$$

$$(2\text{D}) = W M_{2D}(E)$$

$$= W g_v \frac{\sqrt{2m^*(E - E_c)}}{\pi\hbar} H(E - E_c), \qquad (4.30)$$

$$(3\text{D}) = A M_{3D}(E)$$

$$= A g_v \frac{m^*}{2\pi\hbar^2}(E - E_c)H(E - E_c), \qquad (4.31)$$

where $H(E - E_c)$ represents the Heaviside function applied at the conduction band edge $(E = E_c)$, $g_v$ is the electronic band degeneracy, and $m^*$ is the electron effective mass. In comparing the phonon (Eq. (4.13)) and electron (Eq. (4.28)) expressions for heat flow $Q$, we find that they differ by a factor 2 (when expressed in the same integral domain, either energy $E$ or frequency $\omega$); this factor is a result of electronic spin degeneracy.

## 4.3   Thermal Conductance

Thermal conductance $G_Q$ is, by definition, the ratio of heat flow rate $(Q)$ to the driving temperature difference $(T_1 - T_2)$. As the temperature difference becomes small, this ratio can be expressed in differential form:

$$G_Q(T) = \frac{Q(T + \delta T/2, T - \delta T/2)}{\delta T}$$

$$(\text{phonons}) = \frac{1}{2\pi} \int_0^\infty M(\omega)\mathcal{T}(\omega)\hbar\omega \frac{\partial f_{BE}^o}{\partial T} d\omega, \tag{4.32}$$

$$(\text{electrons}) = \frac{1}{\pi\hbar} \int_0^\infty M(E)\mathcal{T}(E)\,(E - \mu)\frac{\partial f_{FD}^o}{\partial T} dE. \tag{4.33}$$

The conductance expressions contain much useful information about the spectral distribution of heat flow for a given carrier. Clearly, both expressions depend on the temperature derivative of the equilibrium distributions functions, which were given generically in Eq. (3.6) and are expressed here for each carrier type:

$$\frac{\partial f_{BE}^o}{\partial T} = (f_{BE}^o)^2 e^{\hbar\omega/k_B T} \left(\frac{\hbar\omega}{k_B T^2}\right), \tag{4.34}$$

$$\frac{\partial f_{FD}^o}{\partial T} = (f_{FD}^o)^2 e^{(E_i - \mu)/k_B T} \left(\frac{E_i - \mu}{k_B T^2}\right). \tag{4.35}$$

These functions are plotted against energy in Figs. 4.3 and 4.4. Two characteristics stand out in comparing the phonon (a) and electron (b) distributions. Firstly, the magnitude of the phonon function is much larger than that for electrons. The reason for this difference is that bosons continue to fill a given state as temperature increases, whereas for electrons, $f_{FD}^o$ is limited to a maximum of 1 by the Pauli exclusion principle. Therefore, the variation of the distribution function is also limited. Secondly, the derivative is always positive for phonons, whereas it transitions from negative to positive at $E = \mu$ for electrons. The reason is that the chemical potential $\mu$ is zero for bosons, and therefore the scaled energy $(\chi)$ must always be positive. For electrons, carriers with energy below $\mu$ are promoted to higher energies as temperature increases. Consequently, the Fermi-Dirac distribution function decreases with increasing temperature for $E < \mu$, and increases for $E > \mu$. However, the integrand in the conductance expression of Eq. (4.33) is always positive because of the multiplicative factor $(E - \mu)$.

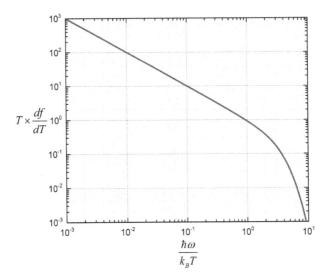

Fig. 4.3   Derivative of the distribution function normalized by temperature $T \times \left( \frac{\partial f_i^o}{\partial T} \right)$ as a function of normalized energy for phonons ($\chi = \frac{\hbar\omega}{k_B T}$).

## 4.4   Spectral Conductance

The integrands of Eq. (4.32) (phonons) and Eq. (4.33) (electrons) are denoted as $G'_Q$ and contain the spectral distribution of energy states that contribute to heat conduction:

$$\text{(phonons)} \ G'_Q(\omega, T) = \frac{1}{2\pi} M(\omega) \mathcal{T}(\omega) \hbar\omega \frac{\partial f_{BE}^o}{\partial T}, \qquad (4.36)$$

$$\text{(electrons)} \ G'_Q(E, T) = \frac{1}{\pi\hbar} M(E) \mathcal{T}(E) (E - \mu) \frac{\partial f_{FD}^o}{\partial T}. \qquad (4.37)$$

If for simplicity we normalize $G'_Q$ by the number of modes $M$ (it generally depends on the dispersion relation through the density of states) and assume the case of perfect transmission ($\mathcal{T} = 1$), then we can define a normalized spectral conductance as:

$$\tilde{G}'_Q = \frac{G'_Q}{C_0 k_B M \mathcal{T}}$$

$$= \boxed{(f_i^o)^2 e^\chi \chi^2}, \qquad (4.38)$$

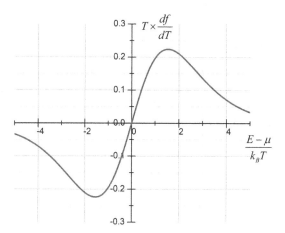

Fig. 4.4 Derivative of the distribution function normalized by temperature $T \times \left( \frac{\partial f_i^o}{\partial T} \right)$ as a function of normalized energy for electrons ($\chi = \frac{E-\mu}{k_B T}$).

$$\text{where } C_0 = (2\pi)^{-1} \text{ (phonons)}$$

$$= (\pi\hbar)^{-1} \text{ (electrons)},$$

$$\text{and } \chi = \frac{\hbar\omega}{k_B T} \text{ (phonons)}$$

$$= \frac{E-\mu}{k_B T} \text{ (electrons)}.$$

Plots of the normalized spectral conductance as a function of the scaled energy $\chi$ (Fig. 4.5 and 4.6) reveal several important characteristics. First, the normalized conductance per mode (recall the $M$ in the denominator of Eq. (4.38)) asymptotes to a value of unity for low-frequency phonons (Fig. 4.5). This result indicates that each mode contributes roughly equally to phonon heat conduction until the phonon energy increases beyond the 'thermal energy' $k_B T$. In contrast, the low-energy electron modes contribute little to heat conduction (Fig. 4.6) because the Pauli exclusion principle embodied in their Fermi-Dirac statistics effectively freezes out their participation. In other words, the thermal energy is not high enough to pro-

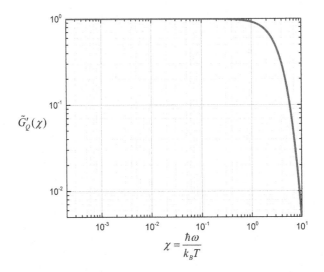

Fig. 4.5    Normalized phonon spectral conductance $\tilde{G}'_Q$ as a function of normalized energy $\chi$.

mote low-energy electrons to available energy states, i.e., near the Fermi energy $(E \approx \mu)$. Therefore, thermal perturbations of electrons away from equilibrium are compensated by a reshuffling of electron occupation near the Fermi energy, where available states exist.

The results for spectral conductance to this point have remained general because they have avoided the need to specify particular dispersion relations for phonons $[\omega(K)$ or $K(\omega)]$ and electrons $(E(k))$. However, these relations are required to determine spectral peaks, which are useful in identifying portions of the spectrum that contribute most substantially to heat conduction. To maintain some of the generality of the foregoing results, here we consider generic frequency (energy) moments of the normalized spectral conductance for phonons (electrons).

The rationale for this approach derives from the factors that comprise the definition of the number of modes $M$, Eqs. (4.23)–(4.25) for phonons and Eqs. (4.29)–(4.31) for electrons. Inspection of the phonon equations reveals that the group velocity $v_g(\omega)$ cancels in each dimensionality. Consequently, the number of phonon modes generally scales as:

$$M(\omega) \sim K(\omega)^{d-1}, \tag{4.39}$$

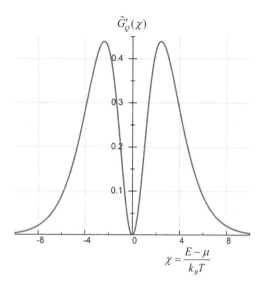

Fig. 4.6 Normalized electron spectral conductance $\tilde{G}'_Q$ as a function of normalized energy $\chi$.

where $d$ is the dimensionality. Thus far, we have not yet invoked a phonon dispersion assumption. The Debye model is surely the most common dispersion approximation for acoustic phonons, with $K \sim \omega$ (see Eq. (3.7)). Consequently, the number of modes scales as:

$$M(\omega) \sim \chi^{d-1}, \qquad (4.40)$$

where $\chi = \hbar\omega/(k_B T)$. Therefore, the quantity $\chi^\alpha \tilde{G}'_Q$ will identify the peaking behavior that we seek. Here, the notional relationship between the exponent $\alpha$ and the dimensionality $d$ is $\alpha \approx (d-1)$. However, a cautionary note is required, as some nanoscale materials such as graphene contain unique phonon branches that do not conform to the $K \sim \omega$ proportionality of the Debye approximation.[3] For now, we will consider $\alpha$ as a general parametric variable.

Figure 4.7 shows the resulting variation of the product of $\chi^\alpha$ and the normalized phonon spectral conductance $\tilde{G}'_Q$ as a function of scaled energy

---

[3]In addition, any dispersion relation $\omega \sim K^x$ that goes through the origin of a $\omega - K$ plot is not appropriate for optical phonons.

Table 4.1    Values of $\chi$ for phonons corresponding to the peaks in $\chi^{\alpha} \tilde{G}'_Q$.

| $\alpha$ | $\chi_{\text{peak}}$ | Notes |
|---|---|---|
| 0 | - | 1D materials under the Debye approximation. No peak. |
| 1 | 2.58 | 2D materials under the Debye approximation. |
| 2 | 3.83 | 3D materials under the Debye approximation. |
| 3 | 4.93 | Special case. |

$\chi = \hbar\omega/k_B T$. All curves except $\alpha = 0$ exhibit peaks for values of $\chi$ in the range of 2 to 5, as shown in Table 4.1. These peaks correspond to a maximization of spectral conductance for a given $\alpha$ as discussed above, and conform to the order of magnitude relation $\hbar\omega_{\text{peak}} \sim k_B T$. The peaks generally increase with increasing $\alpha$ because more energy is contained in higher frequencies (wavevectors) as dimensionality increases, as indicated by the earlier expressions for phonon density of states (Eqs. (2.19)–(2.21)). The absence of a peak for $\alpha = 0$ is related to the discussion associated with Fig. 4.5, which indicates that each mode contributes equally to conductance at low frequencies. The condition $\alpha = 0$ corresponds to the independence of the number of modes $M$ with respect to frequency (or energy), and therefore no peak exists for this special case.

The parallel analysis for electrons is not nearly as straightforward because of the presence of the conduction band minimum energy $E_C$ in the definition of $M(E)$ (see Eqs. (4.29)–(4.31)) and the generally non-zero chemical potential $\mu$. The electronic number of mode results, which derive from a parabolic band assumption, indicate that the number of electron modes scales with dimensionality $d$ as $M \sim E^{(d-1)/2}$. Given the foregoing complications, we illustrate the results for normalized conductance through a contrived condition of $E_C = \mu = 0$. A practical basis for this contrivance is the case of a semiconductor with n-type doping such that the Fermi energy matches the bottom of the conduction band. With these assumptions, the product $\chi^{\eta} \tilde{G}'_Q$ with $\eta = (d-1)/2$ represents the dimensionless conductance with appropriate weighting of the number of modes. The resulting spectral variation appears in Fig. 4.8, and as for phonons the spectral energy peaks occur at small multiples of the thermal energy $k_B T$. The peak values are provided in Table 4.2, and again, they increase mildly with increasing dimensionality according to the dependency of the density of states.

To conclude this section, we highlight other useful spectral peaks. Often, we are interested in the wavelength at peak conductance than frequency or energy. For electrons in metals, the peak conductance occurs near the Fermi energy, and therefore the Fermi wavevector (Eq. (1.39)) can be used

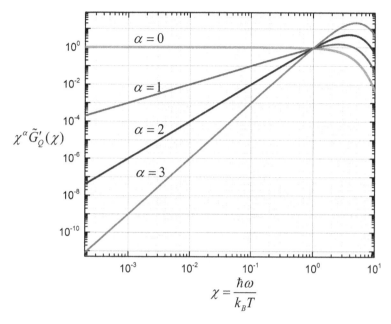

$$\chi = \frac{\hbar \omega}{k_B T}$$

Fig. 4.7 $\chi^\alpha$ moments ($\alpha = 0, 1, 2, 3$) of the normalized phonon spectral conductance $\tilde{G}'_Q$ as a function of normalized phonon energy (or frequency) $\chi$.

Table 4.2 Values of $\chi$ for electrons corresponding to the peaks in $\chi^\eta \tilde{G}'_Q$ for $E_C = \mu = 0$.

| $\eta$ | $\chi_{\text{peak}}$ | Notes |
|---|---|---|
| 0 | 2.40 | 1D materials with parabolic bands. |
| 0.5 | 2.82 | 2D materials with parabolic bands. |
| 1.0 | 3.24 | 3D materials with parabolic bands. |

in the approximation $\lambda_{\text{peak}} \approx 2\pi/k_F$. Taking aluminum as an example, the Fermi energy is $E_F = 11.6$ eV, which dictates a Fermi wavevector of $k_F = 1.745 \times 10^{10}$ m$^{-1}$. The corresponding wavelength is $\lambda_F = 2\pi/k_F = 0.36$ nm, approximately the length of a typical interatomic bond. The minor energy correction of order $k_B T$ as implied by Fig. 4.8 would have a negligible effect on this result.

For phonons, the following example illustrates the calculation of a peak wavelength that maximizes spectral conductance in the frequency spectrum. For bulk (3D) aluminum, the phonon conductance will peak at:

$$\omega_{\text{peak}} = 3.83 \frac{k_B T}{\hbar}, \qquad (4.41)$$

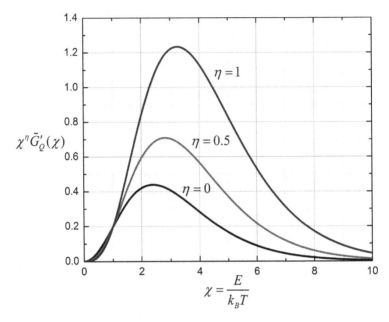

Fig. 4.8   $\chi^{\eta}$ moments ($\eta = 0, 0.5, 1.0$) of the normalized electron spectral conductance $\tilde{G}'_Q$ as a function of normalized energy $\chi$ for the special case of $E_C = \mu = 0$.

which corresponds to $\omega_{\text{peak}} = 38.6 \times 10^{12}$ rad/sec at $T = 77$ K (liquid nitrogen temperature). To find the corresponding approximate wavelength, we start with aluminum's Debye temperature ($\theta_D = 394$ K, Ashcroft and Mermin, 1976) and then use Eq. (3.14) and $\eta_{Al} = 6.03 \times 10^{28}$ atoms/m$^3$ to compute the average group velocity, $v_{g,\text{ave}} = 3375$ m/s. Then, knowing this value, the wavevector at peak conductance is:

$$K_{\text{peak}} = \omega_{\text{peak}}/v_{g,\text{ave}} = 1.14 \times 10^{10} \text{ rad/m}, \qquad (4.42)$$

and the corresponding phonon wavelength at peak conductance is:[4]

$$\lambda_{\text{peak}} = \frac{2\pi}{K_{\text{peak}}} = 0.55 \text{ nm}, \qquad (4.43)$$

(cf., the average interatomic spacing of Al, $\eta_{Al}^{-1/3} = 0.26$ nm).

---

[4]This result is the wavelength at peak conductance in the frequency spectrum and would be slightly different if the problem were posed within the wavelength spectrum, as done to derive Wien's displacement law for thermal radiation—see Fig. 2.7(b).

This type of spectral analysis is particularly important in determining a material's functional dimensionality (i.e., 1D, 2D, or 3D). Considering the example of Al above, if the material is formed in the shape of a nanowire with a square cross-section of 10 nm by 10 nm, then its phonon behavior would be considered bulk (3D) at $T = 77$ K because $\lambda_{\text{peak}} \ll 10$ nm. However, for a lower temperature of $T = 1$ K, the corresponding wavelength at peak spectral (frequency) conductance is $\lambda_{\text{peak}} = 42$ nm $\gg 10$ nm, and the material would essentially behave as a 1D thermal conductor with infinitesimal transverse-direction wavevector components (i.e., $M = 1$) that correspond to rigid-body motion (i.e., infinite corresponding wavelength components).

To close this section, we expand further on the importance of other spectral bases. In all the foregoing development, we have chosen frequency as the spectral quantity for phonon thermal conductance $G'_Q(\omega, T)$ (see Eq. (4.36)) and energy for electronic thermal conductance $G'_Q(\omega, T)$ (see Eq. (4.37)), as well as their normalized counterpart $\tilde{G}'_Q$ (see Eq. (4.38)). Further, in the immediately preceding development, we have calculated the energies, frequencies, and carrier wavelengths corresponding to peaks in these conductances. However, a wavevector or wavelength spectral basis instead of frequency or energy for the conductance would produce slight differences in the peak values of energy, frequency, wavevector, and wavelength. The analogous procedure for thermal radiation produces the well-known Wien's displacement law, which is used to calculate the peak emission wavelength (Modest, 2003). The general conversion from a frequency spectral basis to a wavelength basis can be achieved by applying the chain rule:

$$\tilde{G}'_{Q,\lambda} = \tilde{G}'_Q \frac{d\omega}{dK} \left| \frac{dK}{d\lambda} \right| = \tilde{G}'_Q v_g \frac{2\pi}{\lambda^2}, \tag{4.44}$$

where the absolute value is applied to account for the inverse relationship between frequency and wavelength. Once this procedure is accomplished, the energy (or $\chi$) moments should be applied according to the problem's dimensionality (cf., Fig. 4.7) and with proper accounting for the relationship between energy and wavelength, and then the peak conductance and corresponding wavelength can be calculated. We leave detailed calculations as exercises to interested readers and/or participants in the companion online course.[5]

---

[5]see http://nanohub.org/groups/u/

## 4.5    Example: The Quantum of Thermal Conductance

In this section we derive the quantum of thermal conductance, a seminal result of condensed matter physics. The end result is a simple expression for the maximum rate of heat flow that a single mode can carry between two isothermal reservoirs per unit temperature difference between the reservoirs. This problem was first solved by Rego and Kirczenow (1998). Here, the foundational theory in the preceding sections makes the solution quite straightforward.

The quantum of thermal conductance involves, by definition, a single mode ($M = 1$) and perfect transmission ($\mathcal{T} = 1$), and therefore the preceding 1D results (with $\alpha = \eta = 0$ in Figs. 4.7 and 4.8) apply. For phonons, the corresponding normalized spectral conductance asymptotes to a value of 1 for low frequencies (or scaled energies $\chi = \hbar\omega/k_B T$), and decays to 0 for values of $\chi$ substantially greater than 1. The dimensional quantum of conductance can then be computed through integration of $\tilde{G}'_Q(\chi)$ over $\chi \in (0, \infty)$ as:

$$\hat{G}_{Q,\mathrm{ph}}(T) = \frac{1}{2\pi} \int_0^\infty M(\omega)\mathcal{T}(\omega)\hbar\omega \frac{\partial f_{BE}^o}{\partial T} d\omega$$

$$= \frac{k_B^2 T}{2\pi\hbar} \int_0^\infty M(\chi)\tilde{G}'_Q(\chi)d\chi$$

$$= \frac{k_B^2 T}{2\pi\hbar} \int_0^\infty M(\chi)(f_{BE}^o)^2 e^\chi \chi^2 d\chi, \qquad (4.45)$$

where the terms can be derived from algebraic manipulations of the dimensional conductance (Eq. (4.32)) and normalized spectral conductance (Eq. (4.38)). Based on the graph of $\tilde{G}'_Q(\chi)$ in Fig. 4.5, we expect the integral in Eq. (4.45) (excluding the pre-factor) to have an order of magnitude of 1 assuming that $M = 1$ for all $\chi$, and indeed, the exact integral is $\int_0^\infty (\ldots)d\chi = \pi^2/3$. Therefore, the quantum of thermal conductance can be expressed succinctly as:

$$\boxed{\hat{G}_{Q,\mathrm{ph}}(T) = \frac{\pi k_B^2 T}{6\hbar} \left(= T \times 9.464 \times 10^{-13} \ \mathrm{W/K^2}\right)}. \qquad (4.46)$$

The foregoing result gives the *maximum* phonon conductance that a single mode can provide. The magnitude of overall conductance can only decrease when scattering is present ($\mathcal{T} < 1$), and can only increase by adding more modes $M$ of conductance (e.g., by increasing the conductor's dimensionality). The reason that the conductance quantum increases with temperature is that the carriers in a single mode acquire higher energies (as indicated by the distribution function) as temperature rises. New students are often surprised by the small magnitude of $\hat{G}_Q$ relative to typical thermal conductances found in engineered systems (which are usually $\sim 1$ W/K for cooling technologies). The reconciliation derives from the concept of the number of modes—these 'real' systems exist in higher dimensions that must have tremendously large numbers of modes.

We note here an important caveat in the foregoing development. We have assumed in Eqs. (4.45) and (4.46) that the number of modes remains unity at least through values of $\chi$ at which the phonon occupation has become so small that the integrand is negligible. This assumption allows us to retain the upper limit of integration of $\infty$ so that the integral can be evaluated analytically. However, for sufficiently high temperatures this condition will not be satisfied, and for such cases the proper representation for number of modes will be unity only for values of $\chi$ for which allowable phonon frequencies exists. For example, with the phonon band structure of the 1D diatomic chain (see Fig. 1.18 and the corresponding highlighted bands of Fig. 4.9), the correct conductance expression becomes:

$$G_{Q,\mathrm{ph}}(T) = \frac{k_B^2 T}{2\pi\hbar} \left[ \int_0^{\chi_{\mathrm{max},a}} (f_{BE}^o)^2 e^\chi \chi^2 d\chi + \int_{\chi_{\mathrm{min},o}}^{\chi_{\mathrm{max},o}} (f_{BE}^o)^2 e^\chi \chi^2 d\chi \right].$$
$$(4.47)$$

The foregoing expression will tend toward the value of the quantum of thermal conductance $(\hat{G}_{Q,\mathrm{ph}})$ for temperatures that satisfy $k_B T \ll \hbar\omega_{\mathrm{max},a}$ (or equivalently, $\chi_{\mathrm{max},a} \gg 1$).

The parallel treatment for electrons proceeds similarly, with the electronic quantum of thermal conductance integral expressed as:

$$\hat{G}_{Q,\mathrm{el}}(T) = \frac{k_B^2 T}{\pi\hbar} \int_{\frac{-\mu}{k_B T}}^\infty \tilde{G}_Q'(\chi) d\chi$$

$$= \frac{k_B^2 T}{\pi\hbar} \int_{\frac{-\mu}{k_B T}}^\infty (f_{FD}^o)^2 e^\chi \chi^2 d\chi, \qquad (4.48)$$

where the lower integral bound represents the value of $\chi$ when the electron energy is $E = 0$. Assuming a low thermal energy relative to the chemical

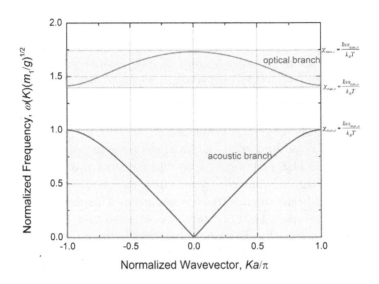

Fig. 4.9   Normalized frequency as a function of normalized wavevector for a diatomic 1D chain with $m_2 = 2m_1$. The shaded regions show the active frequency bands for acoustic and optical branches. The corresponding limits on dimensionless energy $\chi$ are shown on the right side.

potential, the lower bound can be approximated as $-\infty$ to enable analytical evaluation of the integral as $\int_{-\infty}^{\infty} (\ldots) d\chi = \pi^2/3$ (i.e., the same numerical value as the phonon integral). The explicit expression for the electronic quantum of thermal conductance is:

$$\hat{G}_{Q,\text{el}}(T) = \frac{\pi k_B^2 T}{3\hbar} \left( = T \times 1.893 \times 10^{-12} \text{ W/K}^2 \right). \qquad (4.49)$$

We note that the factor of 2 difference between the final results for phonons (Eq. (4.46)) and electrons (Eq. (4.49)) is simply the result of the latter's spin degeneracy.

# Example Problems

---

### Problem 4.1: Application of the Landauer formula

Consider a rectangular block of silicon whose ends are maintained at temperatures of 100 K and 50 K. Assume Debye dispersion for the acoustic branches with group velocities of LA and TA modes being 7200 and 3200 m/s respectively. The group velocities of the optical modes are assumed to be zero (Einstein dispersion). Calculate the heat flux using the Landauer formula assuming ballistic transport, i.e., $\mathcal{T} = 1$. Silicon has a diamond cubic structure with 2 atoms per unit cell and an atom density of $5 \times 10^{22}$ atoms/cm$^3$.

*Solution*

The net heat flux between the hot and cold ends is given by:

$$J_{Q,net} = \sum_p \int_0^\infty \frac{1}{2} \frac{v_{g,p}}{2} D_{3D}(\omega) \hbar\omega \mathcal{T}_p(\omega)[f_{BE}^o(T_1) - f_{BE}^o(T_2)]d\omega.$$

$$(4.50)$$

The density of states $D(\omega)$ under the Debye approximation is given by:

$$D(\omega) = \begin{cases} \frac{\omega^2}{2\pi^2 v_g^3} & \text{if } \omega < \omega_D \\ 0 & \text{if } \omega > \omega_D \end{cases}.$$

$$(4.51)$$

The Debye cutoff wavevector $K_D = (6\pi^2 \eta_a)^{1/3}$ where $\eta_a$ is the unit cell density. Since silicon has two atoms per unit cell, the unit cell density is $2.5 \times 10^{22}$ cells/cm$^3$ (half the atomic density). The Debye cutoff wavevector $K_D = 1.13 \times 10^{10}$ m$^{-1}$. The corresponding Debye cutoff frequency $\omega_D = v_g K_D$ is $8.17 \times 10^{13}$ rad/s for the LA mode and $3.63 \times 10^{13}$ rad/s for the TA mode. Thus the expression in Eq. (4.50) can be written as ($\mathcal{T}_p = 1$):

$$J_{Q,net} = \int_0^{\omega_{D,LA}} \frac{1}{2} \frac{v_{g,LA}}{2} \frac{\omega^2}{2\pi^2 v_{g,LA}^3} \hbar\omega[f_{BE}^o(T_1) - f_{BE}^o(T_2)]d\omega$$

$$+ 2\int_0^{\omega_{D,TA}} \frac{1}{2} \frac{v_{g,TA}}{2} \frac{\omega^2}{2\pi^2 v_{g,TA}^3} \hbar\omega[f_{BE}^o(T_1) - f_{BE}^o(T_2)]d\omega,$$

$$(4.52)$$

where the factor of 2 in the second integral on the right side accounts for two degenerate TA modes. Also note that the optical branches do not contribute to heat flux as the group velocity is assumed to be zero. Substituting $T_1 = 100$ K and $T_2 = 50$ K and evaluating the integrals numerically, we obtain $J_{Q,\text{net}} = 1.85 \times 10^{10}$ W/m$^2$.

---

**Problem 4.2: Number of modes**

Consider an aluminum block of square cross-section (1cm×1cm). Calculate the phonon frequency for a single acoustic branch at which the spectral conductance is a maximum for a temperature of 300 K. Also obtain the number of modes $M(\omega)$ at this frequency. Assume Debye dispersion with a group velocity of 3400 m/s.

*Solution*

For 3D materials under the Debye approximation, spectral conductance is a maximum when $\hbar\omega/k_B T = 3.83$ (see Table 4.1). At $T = 300$ K, this corresponds to a phonon angular frequency of $1.504 \times 10^{14}$ rad/s. The number of modes $M(\omega)$ is given by:

$$M(\omega) = A\frac{\omega^2}{4\pi v_g^2}. \tag{4.53}$$

Substituting for the cross-sectional area $A = 0.01$ m$^2$ and $v_g = 3400$ m/s, we obtain $M(\omega = \omega_{\text{peak}}) = 1.56 \times 10^{18}$.

---

**Problem 4.3: Quantum of thermal conductance**

Consider an experiment to measure the quantum of thermal conductance on a monoatomic 1D chain with $g = 25$ N/m and $m = 28$ amu. Assume ideal coupling of the 1D chain with hot and cold reservoirs ($\mathcal{T} = 1$). Calculate the maximum temperature at which the experiment needs to be performed so as to measure a conductance within 10% of the quantum of conductance.

*Solution*

The phonon thermal conductance is given by the following integral:

$$G_Q(T) = \frac{k_B^2 T}{2\pi\hbar} \int_0^\infty M(\chi)\tilde{G}_Q'(\chi)d\chi$$

$$= \frac{k_B^2 T}{2\pi\hbar} \int_0^\infty M(\chi)f_{BE}^o{}^2 e^\chi \chi^2 d\chi$$

$$= \frac{k_B^2 T}{2\pi\hbar} \int_0^{\frac{\hbar\omega_{max}}{k_B T}} f_{BE}^o{}^2 e^\chi \chi^2 d\chi, \qquad (4.54)$$

where $\omega_{max} = 2\sqrt{g/m}$ is the maximum phonon frequency of the 1D chain. The last equality is obtained from the fact that the number of modes is 1 only for $\omega < \omega_{max}$. The density of states and the number of modes is zero for $\omega > \omega_{max}$. In order to measure a conductance within 90% of the quantum of conductance, we need the upper limit of the integral to be greater than 4.7, i.e.,

$$\int_0^{4.7} f_{BE}^o{}^2 e^\chi \chi^2 d\chi \approx 0.9 \times \int_0^\infty f_{BE}^o{}^2 e^\chi \chi^2 d\chi. \qquad (4.55)$$

The maximum phonon frequency of the 1D chain for the given set of parameters is $4.63\times10^{13}$ rad/s. From the equation $\hbar\omega_{max}/k_B T = 4.7$, we obtain a maximum temperature of approximately 75 K. Thus the experiment needs to be performed at a temperature less than 75 K so as to measure a conductance within 10% of the quantum of thermal conductance. Figure 4.10 shows the variation in the ratio of conductance $(G)$ to the quantum of thermal conductance $(G_o)$ as a function of temperature. The ratio is close to 1 for very low temperatures and decays for temperatures greater than 75 K.

*Thermal Energy at the Nanoscale*

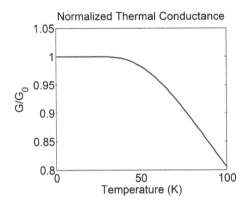

Fig. 4.10   Variation of conductance normalized by the quantum of conductance as a function of temperature.

---

### *Problem 4.4: Spectral thermal conductance*

(a) Obtain the scaling relation between the number of modes $M(\omega)$ and the non-dimensional energy $\chi = \hbar\omega/k_B T$. Assume that the dispersion relation is given by $\omega \sim K^n$ and $d$ is the number of dimensions.

(b) The online Chapter 4 CDF tool[6] plots the $\chi^\alpha$ moments of the normalized spectral conductance $\tilde{G}'_Q(\chi)$ where $\alpha$ is a function of $d$ and $n$ obtained in part (a) of this problem. The CDF tool allows the user to specify the dimension $d$ and the exponent $n$ in the dispersion relation. Use the tool to observe the trend in variation of $\chi_{\text{peak}}$ (the normalized energy at which $\chi^\alpha \tilde{G}'_Q(\chi)$ is a maximum). Also provide a physical explanation for the observed trend.

(c) The LA and TA branches of graphene can be approximated with Debye dispersion relations where the group velocities are given by $v_g(LA) = 21300$ m/s and $v_g(TA) = 13600$ m/s. The ZA branch is approximated by a quadratic dispersion relation near the Brillouin zone center (see Appendix) of the form $\omega = CK^2$, where $C = 5 \times 10^{-7}$ m$^2$/s. At $T = 300$ K, use the CDF tool to calculate the phonon wavelengths at which the spectral conductance is a maximum for the LA, TA and ZA branches.

---

[6]See http://nanohub.org/groups/cdf_tools_thermal_energy_course/wiki

*Solution*

(a) Number of modes $M \sim K^{d-1} \sim \omega^{(d-1)/n} \sim \chi^{(d-1)/n}$.

(b) From the online Chapter 4 CDF tool, the peak frequency increases as the number of dimensions increases for a given exponent $n$ in the dispersion relation. This result is a consequence of the fact that more energy is concentrated in high frequency phonons as dimensionality increases. Also no peak exists in the spectral conductance for the 1D case.

(c) For the 2D case and under the Debye approximation ($n = 1$), we obtain from the CDF tool $\chi_{\text{peak}} \approx 2.58$ (see Fig. 4.11). When quadratic dispersion ($n = 2$) is considered, we obtain $\chi_{\text{peak}} \approx 1.8$ (see Fig. 4.11). At $T = 300$ K, these correspond to angular frequencies of 101.33 THz ($n = 1$) and 70.7 THz ($n = 2$). Using the dispersion relation, we then obtain $\lambda_{\text{peak}} = 1.3$ nm for the LA branch, $\lambda_{\text{peak}} = 0.84$ nm for the TA branch and $\lambda_{\text{peak}} = 0.53$ nm for the ZA branch.

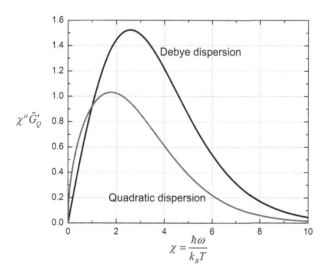

Fig. 4.11 Variation of spectral conductance ($\chi^{\alpha} \tilde{G}'_Q$) as a function of normalized frequency ($\chi$) for $d = 2$. The black curve corresponds to Debye dispersion ($n = 1$) and the red curve corresponds to quadratic dispersion ($n = 2$).

Chapter 5

# Carrier Scattering and Transmission

The content in this chapter seeks to move toward closure of the contact-device-contact problem originally introduced in Fig. 1.1. Looking back at the overarching questions posed in that section, we have answered each of them to varying degrees of depth, except the last two:

- How do the carriers scatter as they move through the material?
- How do the boundaries and interfaces impede carriers?

Answers to these questions are addressed in this chapter with the caveat that, like many expositions on technical subjects, this text has deferred the most complicated subjects toward its conclusion. Perhaps this tactic is employed so commonly in order to minimize early attrition amongst the readership. Here, we might hope that the preceding topics and their elucidation have motivated the reader to take on more challenging topics in the interest of obtaining full understanding, if not mastery, of thermal energy transport in our motivating model problem. Moreover, this chapter focuses (although not exclusively) on phonon scattering, in deference to the many excellent sources on electron scattering (Lundstrom, 2009).

## 5.1 Scattering Analysis in the Landauer Formalism

We begin with a general expression for the spectral thermal conductance without regard to carrier type (Eq. (4.36) for phonons; Eq. (4.37) for electrons):

$$G'_Q = C_0 M(E) \mathcal{T}(E)(E - \mu)\frac{\partial f_i^o}{\partial T},$$ 
(5.1)

where $C$ is a constant that depends on carrier type. The purpose here is to develop some intuition about the transmission function $\mathcal{T}$. With reference to the device region of Fig. 1.1, we might reasonably expect the transmission function to decrease with increasing device length $L$ and with a decrease in the distance that a carrier travels between 'scattering' events (i.e., the more often a carrier collides with something, the less likely it is to reach the other side). We will denote this scattering length as $\Lambda$, which is also called the mean free path. Our rudimentary intuitive model would then suggest:

$$\mathcal{T} \sim \frac{\Lambda}{L}. \tag{5.2}$$

In fact, this intuitive formulation can be quite accurate under 'diffusive' transport (i.e., when many scattering events occur in one traversal of the device). However, the simple model becomes invalid for ballistic (and quasi-ballistic) transport (i.e., $\Lambda > L$) because it produces a transmission value greater than unity. An easy remedy is to modify the denominator slightly:

$$\boxed{\mathcal{T} = \frac{\Lambda}{\Lambda + L}}. \tag{5.3}$$

This expression now satisfies both the diffusive case, for which $\frac{\Lambda}{\Lambda+L} \approx \frac{\Lambda}{L}$ because $\Lambda \ll L$, and the ballistic case ($\Lambda \gg L$) for which $\mathcal{T} = 1$.

This model under diffusive conditions actually allows for a direct comparison to the thermal conductivity expression derived from kinetic theory in Section 3.3, with a result that may be reassuring to readers concerned with the qualitative nature of the foregoing reasoning.

## 5.2 Thermal Conductivity Revisited

Thermal conductivity $\kappa$ is intimately related to thermal conductance $G_Q$, as the root words imply. The concept of thermal conductivity was developed, in general, to represent a property of a material that does not depend on its size or shape. Intuition suggests that thermal conductance, in contrast, should decrease as the path length of heat flow (i.e., $L$) increases, and that it should increase as the cross-sectional 'area' increases. The following relation normalizes these geometric factors out of the conductance to represent thermal conductivity:

$$\kappa = \frac{L}{\text{'area'}} G_Q. \tag{5.4}$$

Then, substituting the integral phonon expression for $G_Q$ into Eq. (5.4), the thermal conductivity for a single phonon branch becomes:

$$\kappa = \frac{L}{\text{`area'}} \frac{1}{2\pi} \int_0^\infty M(\omega) T(\omega) \hbar\omega \frac{\partial f_{BE}^o}{\partial T} d\omega. \tag{5.5}$$

Next, we recall the relation between the number of modes and 'area' (see Eq. (4.16)):

$$M = \text{`area'} \times M_{dD}, \tag{5.6}$$

where $d$ again represents the problem's dimensionality. Then, substitution of the foregoing relation and the diffusive form of the transmission function $(T = \Lambda/L)$ into Eq. (5.5) yields:

$$\kappa = \frac{L}{\text{`area'}} \frac{1}{2\pi} \int_0^\infty \text{`area'} \times M_{dD}(\omega) \frac{\Lambda}{L} \hbar\omega \frac{\partial f_{BE}^o}{\partial T} d\omega$$

$$= \frac{1}{2\pi} \int_0^\infty M_{dD}(\omega)\Lambda(\omega) \hbar\omega \frac{\partial f_{BE}^o}{\partial T} d\omega, \tag{5.7}$$

where the mean free path is shown to allow for frequency dependence generally. The last substitution in Eq. (5.7) originates from the dimension-specific mode densities (see Eqs. (4.17)–(4.19)), which take the general form:

$$M_{dD} = \pi \langle v_{gx} \rangle D_{dD}(\omega), \tag{5.8}$$

where $\langle v_{gx} \rangle$ is the directionally averaged carrier velocity corresponding to the problem's dimensionality (see Eqs. (4.23)–(4.25)). Finally, assuming for simplicity that the scattering length and average velocity do not depend on frequency, the phonon thermal conductivity can be expressed as:

$$\kappa = \frac{1}{2} \langle v_{gx} \rangle \Lambda \underbrace{\int_0^\infty \hbar\omega D_{dD}(\omega) \frac{\partial f_{BE}^o}{\partial T} d\omega}_{c_v \text{ of a phonon branch}}, \tag{5.9}$$

where the underbrace highlights that the integral term is identical to that in the expression for phonon specific heat (Eq. (3.34)) for a single phonon branch; a summation over all branches would give the total thermal conductivity. Importantly the resulting general form of the expression for thermal conductivity $\kappa \sim c_v v_g \Lambda$ conforms to the thermal conductivity expression derived in Chapter 3 from kinetic theory (see Eq. (3.42)). This consistency is particularly reassuring given the rather qualitative intuition on which the diffusive transmission function was postulated $(T = \Lambda/L)$. This result also provides insight into another common thermal property called thermal diffusivity $\alpha$, which is defined as the ratio of thermal conductivity to volumetric specific heat:[1]

$$\alpha \equiv \frac{\kappa}{c_v} = \frac{1}{2}\langle v_{gx}\rangle\Lambda. \tag{5.10}$$

This result is memorable in that the diffusivity is simply proportional to the product of carrier velocity and mean free path.

The same treatment for electrons in which the conductance is expressed in terms of an energy integral would produce a very similar result to that derived here for phonons, and Appendix B provides a qualitative analysis of electronic heat conduction as compared to phonons for graphene. Often, the electronic contribution to thermal conductivity for a given material is inferred from its electrical conductivity through the Wiedemann-Franz Law. The reason for this preference is the relative simplicity of measuring electrical conductivity (using common hardware such as multimeters) as compared to thermal conductivity, which typically requires more exotic calorimeters or indirect methods for very small materials Wang (2012). The Wiedemann-Franz law, which is semi-empirical, is usually expressed in terms of the Lorentz number $L_e$:

$$\kappa_e = \sigma_e L_e T, \tag{5.11}$$

where $\sigma_e$ is the electrical conductivity.

Proper selection of a value for $L_e$ requires some knowledge of the material's band structure. In a companion book within this series, Lundstrom and Jeong (2013) cover this topic in detail, resulting in the following Lorenz number for parabolic bands and an energy-independent mean free path:

$$L_e = C\left(\frac{k_B}{q}\right)^2, \tag{5.12}$$

where $C = 2$ for a non-degenerate band, and $C = \pi^2/3$ for a degenerate band.

## 5.3   Boundary and Defect Scattering

The primary challenge in accurate modeling of thermal conductance and conductivity involves finding a suitable form of the mean free path $\Lambda$. Scattering processes encompass a broad range of disparate physical concepts and

---

[1] The more common form uses the mass-based specific heat, whereby $\alpha = \frac{\kappa}{\rho c_p}$.

processes. They are often assumed to be independent of each other (i.e., the occurrence of one type of scattering event does not affect the other types). Phonon scattering can occur by a variety of mechanisms, and a truly comprehensive exposition is beyond the present scope. Instead, this text covers the most common physical scattering processes and associated models for frequency (energy) and temperature dependence. In this regard, the approach here harkens to that of Ziman (1972, p. 71), who wrote:

> This subject is open-ended leading to such intractable problems as the dynamics of completely disordered systems such as liquids and glasses. Nevertheless, if we confine ourselves to isolated defects ... we can now understand a number of interesting physical phenomena.

For greater depth, the book by Kaviany (2008) admirably describes details of internal phonon scattering process and models, as well as guidelines for 'quilting' the models together over relevant parameter spaces such as frequencies and temperatures.

A common surrogate for $\Lambda$ is the scattering time $\tau$:

$$\tau = \frac{\Lambda}{v_g}. \tag{5.13}$$

This term represents the mean time between scattering events. An equally (if not more) common descriptor is the scattering rate, which is simply the inverse of the scattering time, $\tau^{-1} = v_g/\Lambda$. Only toward the end of this chapter will we assemble the various scattering types into an 'effective' form. The next section begins with the conceptually simplest type—boundary scattering.

### 5.3.1 *Boundaries*

When a carrier encounters a material surface, it reflects back into the the material. These reflections can impede heat flow, but the extent to which this occurs depends on factors such as the surface roughness, and the relative importance of such boundary scattering among all scattering events greatly increases when the surfaces are near to each other, such as in nanowires and thin films. Figure 5.1 shows a Cartesian geometry with confined dimensions $l_1$ and $l_2$.

Intuition suggests that the scattering length should be closely related to the thicknesses of the object. For example, we should expect that $\Lambda \approx l_1$ for 'cross-plane' heat flow (i.e., from bottom to top across the this film). In

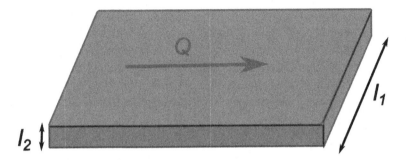

Fig. 5.1   Heat flow through a rectangular cross section of $l_1 \times l_2$. The boundaries can constrain the carrier mean free path.

this case, the boundary scattering rate would be:

$$\tau_b^{-1} \approx Cv_g/\Lambda = Cv_g/l_1, \tag{5.14}$$

where the constant $C$ should be approximately $1/3$ based on the arguments in the derivation of kinetic theory in Chapter 3. However, a cautionary note is in order here because if the boundaries do indeed dominate transport, then the medium should be considered ballistic (or quasi-ballistic) rather than diffusive. Moreover, if this scattering rate is to be meaningful in the context of thermal conductivity, then the upper and lower surfaces would need to be connected to contact materials, and the boundaries would be best represented as interfaces—a subject deferred to Section 5.6.

Conversely, for 'in-plane' heat flow in the direction shown in Fig. 5.1, we expect many scattering events within the material (i.e., the diffusive approximation should apply well). However, the geometric interpretation of the mean free path is not as straightforward as for the cross-plane case. For such a relationship, Holland (1963) suggested:

$$\tau_b^{-1} = \frac{v_g}{F\mathcal{L}}, \tag{5.15}$$

where

$$\mathcal{L} = 2\sqrt{\frac{l_1 l_2}{\pi}}. \tag{5.16}$$

The factor $F$ in Eq. (5.16) is a fitting factor that depends, in part, on the roughness of the surfaces. Surface roughness is important because it dictates whether the scattering is 'specular' (i.e., mirror-like) or 'diffuse'. The distinction is crucial because specular reflection will preserve carrier momentum in the direction of interest, whereas diffuse reflection will randomize the scattered direction and therefore impede heat flow (Berman *et al.*, 1953).

The factor $F$ is normally considered a constant that is "adjusted to give an exact fit at low temperature" (Holland, 1963). However, its possible dependence on surface roughness deserves some scrutiny because the degree of roughness must be considered in relation to the carrier wavelength. For example a boundary with a characteristic surface roughness of 100 nm will appear quite smooth to a carrier with a wavelength of $\lambda = 10^3$ nm, whereas it will seem very rough to a carrier with a wavelength of $\lambda = 1$ nm, as illustrated in Fig. 5.2. This disparity would imply a wavelength (or frequency, i.e., energy) dependence for this factor, and indeed, a wavelength-dependent 'specularity parameter' can be defined (Yang and Chen, 2004). This subject remains in the realm of contemporary research, and its various permutations are not included here for the sake of clarity and brevity.

### 5.3.2 *Defects*

A common and analytically accessible scattering type involves point defects such as substitutional impurity atoms, which are foreign atoms that occupy a regular lattice site. As suggested notionally in Fig. 5.3, these defects have a small radius of influence (typically about one bond length). The resultant discontinuity in the lattice properties (i.e., atomic mass and

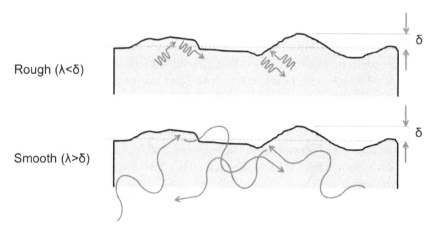

Fig. 5.2   Boundary scattering illustration showing that the same physical boundary can appear to be rough (top) to small wavelengths and smooth (bottom) to large wavelengths.

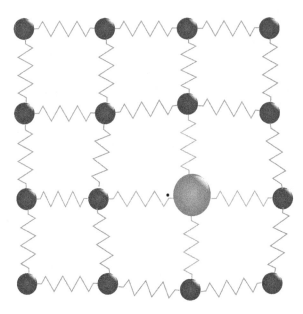

Fig. 5.3    Schematic of a point defect in a lattice. In this case, the defect is termed substitutional because it sits at a regular lattice site. The defect alters the local bonds, thereby creating an extended cross-section of its influence.

spring constants) will cause a wave to alter direction, or scatter. These processes tend to preserve the wave's energy and therefore are termed *elastic scattering* events, which are characterized by:

$$\omega = \omega' \quad \text{(elastic scattering)}, \tag{5.17}$$

where $\omega'$ is the phonon frequency after the scattering event.

The basic point-defect scattering process can be analyzed using the conceptual framework shown in Fig. 5.4. The circular tube contains a distribution of fixed scattering sites, each of diameter $d$. A carrier proceeding through a tube of diameter $2d$ is likely to be scattered by the sites. If the volumetric concentration of scattering sites is $n_i$ throughout the domain, then the number of scatterers in the tube is $n_i \pi d^2 L$. The estimated mean-free path is then expected to be the ratio of the tube length $L$ to the number of scatterers:

$$\Lambda_i = \frac{L}{n_i \pi d^2 L} = \frac{1}{n_i \pi d^2}. \tag{5.18}$$

In practice, the scattering site diameter is not easily derived. Instead, the concept of scattering cross-section $\sigma$ is used to replace the $\pi d^2$ term

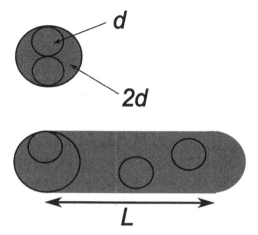

Fig. 5.4    Schematic of a scattering tube that is aligned with a direction of energy transport. Defects within the tube each have a diameter of $d$, making the effective diameter of the scattering tube $2d$.

of Eq. (5.18). Then, using the relation $\tau^{-1} = v_g/\Lambda$, the scattering rate becomes:

$$\tau_i^{-1} = \alpha \sigma n_i v_g, \tag{5.19}$$

where $\alpha$ is a constant of the order of 1 that is commonly obtained from curve-fitting to property data.

As a special case, we consider Rayleigh scattering, which describes point defect scattering for cases when the defect diameter $d$ is much smaller than the phonon wavelength $\lambda$, i.e., $d \ll \lambda$. The defining characteristic of Rayleigh scattering is an inverse fourth-power dependence of the scattering cross section on wavelength:

$$\sigma \sim \frac{1}{\lambda^4}. \tag{5.20}$$

In effect, very long waves more readily pass through these point defects without scattering than shorter wavelengths (although all must be much larger than the defect size for the Rayleigh scattering model to apply).

Klemens (1951) showed that this type of scattering for long phonon waves (for which the Debye approximation is valid, $\omega \sim \lambda^{-1}$) produces the following scattering rate:

$$\tau_{i,R}^{-1} = \frac{n_i V^2 \Delta m^2}{4\pi v_g^3 m^2} \omega^4, \tag{5.21}$$

where $m$ and $V$ are the mass of the host atom and primitive cell volume, respectively; $n_i$ is the impurity volumetric concentration; and $\Delta m$ is the difference in mass between the impurity and host atoms.

## 5.4    Phonon-Phonon Scattering Fundamentals

Phonons in a perfectly harmonic crystal will propagate independently, without being influenced by each other for the simple reason the each bond will act with the same spring constant regardless of the instantaneous position of any atom. Therefore, local atomic displacements will not affect the propagation of other waves by perturbing the effect of the bond-spring. However, real materials—even 'perfect' crystals—possess anharmonicity in their bonds as shown in Fig. 5.5 that causes wave-wave scattering.

The most common intra-phonon scattering events are the so-called 'three-phonon' processes in which either two phonons combine into one, or one phonon decomposes into two. Of course, many other combinations involving more than three phonons are possible, but as we will find, the rules that govern the allowability of such multi-phonon processes are strict,

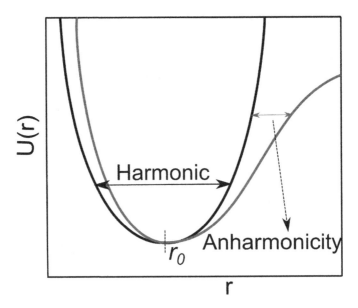

Fig. 5.5    Bond energy diagram showing the ideal harmonic behavior and the real anharmonic shape of the potential energy curve.

# 3-phonon interactions

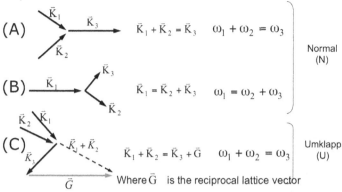

Fig. 5.6 Three-phonon scattering processes of types A (2 in, 1 out, with momentum conservation), B (1 in, 2 out, with momentum conservation), and C (2 in, 1 out, without momentum conservation).

rendering the three-phonon processes most likely. Figure 5.6 illustrates three major types of three-phonon scattering processes:

**A:** Two incoming phonons scattering into a third while conserving energy and momentum

**B:** One phonon decaying into two outgoing phonons while conserving energy and momentum

**C:** Two incoming phonons scattering into a third while conserving energy *but not* momentum

The crucial distinction among these scattering types involves the treatment of momentum. Types A and B, which are termed 'Normal' or just 'N' processes, both conserve momentum, whereas type C does not. Consequently, N processes do not directly impede heat conduction because the net carrier momentum in a direction of interest remains unaffected by scattering.

The C type of scattering is termed an 'Umklapp' or simply 'U' process[2] and refers to a reversal or 'folding' of the phonon wavevector (and thus momentum, $\vec{p} = \hbar \vec{K}$) back into the first Brillouin zone. This reversal is required because phonons with wavevectors outside the first Brillouin zone

---

[2]The term *Umklapp* is of German origin and means 'to be folded'.

have corresponding wavelengths that are simply too small to be supported by the lattice. The remedy is to translate the wavevector by a harmonic translation, i.e., the reciprocal lattice vector $\vec{G}$ that returns the resultant to the first Brillouin zone:

$$\vec{K}_1 + \vec{K}_2 = \vec{K}_3 + \vec{G}. \tag{5.22}$$

A k-space sketch of N (left) and U (right) processes for graphene is shown in Fig. 5.7. We note that all three-phonon U processes involve two incoming phonons and one outgoing phonon—akin to the type A Normal process. A U process comparable to the type B Normal process—decay of a single incoming phonon—does not exist because such a phonon would not be allowed outside the first Brillouin zone.

Of course, both types of processes—N and U—must satisfy energy conservation:

$$\omega_1 + \omega_2 = \omega_3 \quad \text{or} \quad \omega_1 = \omega_2 + \omega_3. \tag{5.23}$$

This energy conservation requirement proves to be very restrictive, allowing only a small set of three-phonon combinations to occur. While this topic is beyond the scope of the present exposition, we include here the

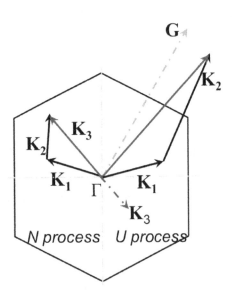

Fig. 5.7 **k**-space diagram of graphene showing example N (left) and U (right) phonon scattering processes.

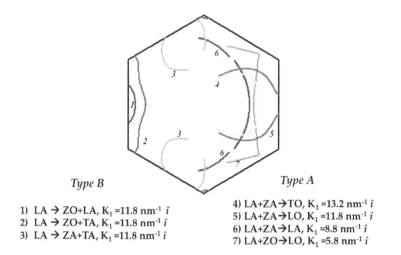

Type B

1) LA → ZO+LA, $K_1$ =11.8 nm$^{-1}$ $\hat{\imath}$
2) LA → ZO+TA, $K_1$ =11.8 nm$^{-1}$ $\hat{\imath}$
3) LA → ZA+TA, $K_1$ =11.8 nm$^{-1}$ $\hat{\imath}$

Type A

4) LA+ZA→TO, $K_1$ =13.2 nm$^{-1}$ $\hat{\imath}$
5) LA+ZA→LO, $K_1$ =11.8 nm$^{-1}$ $\hat{\imath}$
6) LA+ZA→LA, $K_1$ =8.8 nm$^{-1}$ $\hat{\imath}$
7) LA+ZO→LO, $K_1$ =5.8 nm$^{-1}$ $\hat{\imath}$

Fig. 5.8   k-space of graphene showing the highly restricted modes that satisfy the scattering selection rules for various three-phonon scattering examples. Based on the model reported by Singh *et al.* (2011b).

main principles. First, an example of the allowed phonon combinations for specific types of graphene scattering are indicated in Fig. 5.8, which shows lines on which energy balance can be achieved within the $\vec{K}_2$ k-space for specific $\vec{K}_1$ vectors oriented in the $x$-direction (Singh *et al.*, 2011b). The results reveal that only a small subset of k-space can participate in scattering with the fixed incident phonon.

Because of this complication, exact analytical models for phonon-phonon scattering are generally not possible. The fully rigorous recourse is to employ quantum theory with perturbations that account for anharmonicity to calculate so-called interaction matrices between incoming and outgoing phonons through a process known as Fermi's golden rule. We exclude the details of this approach from the present text; Kaviany (2008, Appendix E) provides a full derivation and numerous examples.

Instead, we explain here some phenomenological and qualitative features of phonon-phonon scattering rate models. The two most important such principles follow:

- The scattering rate should generally increase with increasing frequency, because the phonon density of states increases with frequency.

- The scattering rate should generally increase with temperature, because more high-frequency phonons are populated and thus available to be scattered.

The foregoing principles combine into a general phenomenological scattering rate model:

$$\tau_{p-p}^{-1} = B\omega^n T^m, \tag{5.24}$$

where $B$ is a constant, and $n$ and $m$ are typically integers greater than or equal to one. Klemens (1958) provided an early comprehensive review for bulk crystals, and the book by Kaviany (2008) contains a thorough analysis and demarcation of the applicability of various models, particularly for bulk materials. We refrain here from a repetition of these quality sources, to which the interested reader is referred.

## 5.5   The Effective Scattering Rate

Once individual scattering models have been established, they need to be combined into an effective scattering term. Intuition might suggest that an averaging process could be defined, but the term 'mean' in 'mean free path' suggests that an averaging process has already been applied for individual scattering processes. Instead, an additive approach is commonly used that sums the various scattering rates through what is termed the 'relaxation time approximation' (RTA) when used in the context of the Boltzmann transport equation (BTE) and its derivatives (Joshi and Majumdar, 1993). The summation is performed using the so-called Matthiessen's Rule:

$$\tau_{\text{eff}}^{-1} = \sum_{\text{scat. proc. } j} \tau_j^{-1}. \tag{5.25}$$

However, some important caveats apply. First, the foregoing 'rule' assumes that one type of scattering process in the summation does not affect any of the others. In other words, a defect scattering process does not make a boundary scattering event more likely or less likely to occur. Second, only the scattering types that impede heat flow should be included in the calculation of $\tau_{\text{eff}}^{-1}$. To understand why, recall the earlier expression for thermal conductivity from kinetic theory:

$$\kappa = \frac{1}{3}v_g \Lambda_{\text{eff}} c_v = \frac{1}{3}v_g^2 \tau_{\text{eff}} c_v. \tag{5.26}$$

Consequently, the N processes for phonon-phonon scattering discussed in Section 5.4 should not be included in Matthiessen's rule. However, they do affect U processes by increasing the number of populated high-K phonons, which in turn are more likely to participate in U processes. Therefore, even though N processes do not directly impede heat flow, they are important because they 'feed' U processes. Callaway (1959) provides a useful model for this N/U interaction, and Ni and Murthy (2012) demonstrate how it can be applied to contemporary research.

As a simple example, we consider a 1D material ($M = 1$) with a constant defect scattering rate and a power-law U process scattering rate:

$$\tau_i^{-1} = 10^{10} \text{ s}^{-1}, \tag{5.27}$$

$$\tau_{U,p-p}^{-1} = \left(10^{-17} \text{ s/K}\right) \omega^2 T. \tag{5.28}$$

The effective scattering rate becomes:

$$\tau_{\text{eff}}^{-1} = 10^{10} \text{ s}^{-1} + \left(10^{-17} \text{ s/K}\right) \omega^2 T. \tag{5.29}$$

Figure 5.9 shows effective scattering rates as a function of normalized frequency [$\chi = \hbar\omega/(k_B T)$] for three different temperatures. For all

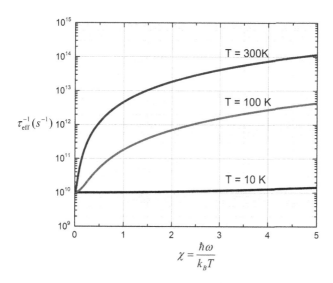

Fig. 5.9 Effective scattering rates as a function of normalized frequency [$\chi = \hbar\omega/(k_B T)$] for three different temperatures, using Matthiessen's rule for the example of combined defect and U process scattering.

temperatures the low-frequency scattering rate is fixed by the defect scattering term. However, for the higher temperatures, the U process scattering rate steadily overtakes the defect rate as frequency increases, leading to a large increase that dominates the effective scattering rate for almost all frequencies. Only at the lowest temperature is defect scattering significant.

The effect of temperature can be further understood through the variation of thermal conductivity as shown in Fig. 5.10 (which assumes a constant group velocity of $v_g = 1000$ m/s for simplicity). The thermal conductivity is calculated using Eq. (5.7):

$$\kappa = \frac{1}{2\pi} \int_0^\infty \Lambda(\omega) M_{dD}(\omega) \hbar\omega \frac{\partial f_{BE}^o}{\partial T} d\omega$$

$$(1D) = \frac{1}{2\pi} \int_0^\infty \frac{v_g}{\tau^{-1}(\omega)} \hbar\omega \frac{\partial f_{BE}^o}{\partial T} d\omega. \tag{5.30}$$

In Fig. 5.10 the thermal conductivity for temperatures below 10 K inherits the characteristic temperature dependence of the specific heat ($c_v \sim T^1$ for

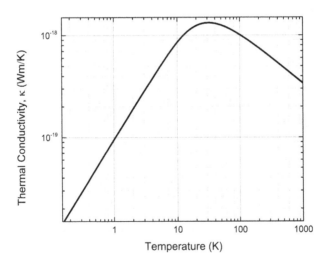

Fig. 5.10   Thermal conductivity of a 1D material as a function of temperature for the example of combined defect and U process scattering. The region below $T = 10$ K is dominated by defect scattering. Above 10 K, U scattering and the plateauing of specific heat with temperature become prominent.

this 1D example). Thereafter, a transition occurs in which the temperature dependence becomes dominated by U process scattering ($\tau_U^{-1} \sim T$) while the specific heat begins to asymptote toward a constant value (according to the Law of Dulong and Petit). As temperature increases further, the scattering rate continues to increase, resulting is a pronounced decrease in thermal conductivity. This simple example illustrates qualitatively the typical regimes of a material whose thermal conductivity is dominated by phonons:

- Increasing thermal conductivity with increasing cryogenic temperatures dominated by the material's specific heat.
- A range of transition temperatures in which the increase in specific heat begins to moderate while phonon-phonon scattering becomes prevalent. The peak thermal conductivity occurs in this region.
- A decreasing thermal conductivity at high temperatures as specific heat becomes constant while the phonon-phonon scattering rate continues to increase.

## 5.6 Interfacial Transmission

Individual nanomaterials can exhibit extreme thermal properties—both high and low magnitudes. In applications such as thermal insulation or thermoelectrics, the objective is to suppress heat conduction, and a common strategy is to introduce numerous heterogeneous material interfaces that reflect thermal energy carriers. At the other extreme, the engineering objective is typically to translate the outstanding thermal properties of *individual* nanoscale elements to more practical human length scales by connecting many (typically billions) such elements to each other and to the 'bulk' contacts of the real world. Achieving such a circumstance is, however, made very difficult because of the many local interfaces whose presence often mutes the very properties for which the nanoscale elements were chosen. In both cases—high and low property extremes—an understanding of heat flow across interfaces is essential.

Interfacial heat flow is often quantified by its *thermal boundary (interface) resistance*[3] or its inverse, the thermal boundary conductance. Using

---

[3] For solid-fluid interfaces, the resulting resistance is often called the Kapitza resistance Kapitza (1941).

our prior framework, the expression for thermal boundary resistance is:

$$R_b = \frac{T_1 - T_2}{Q_{\text{ph}}}$$

$$= \frac{T_1 - T_2}{\sum_p \frac{1}{2\pi} \int_0^\infty \hbar\omega M(\omega)\mathcal{T}(\omega)\left[f^o_{BE}(T_1) - f^o_{BE}(T_2)\right] d\omega} \tag{5.31}$$

$$\approx \left[\sum_p \frac{1}{2\pi} \int_0^\infty \hbar\omega M(\omega)\mathcal{T}(\omega)\frac{\partial f^o_{BE}}{\partial T} d\omega\right]^{-1}, \tag{5.32}$$

where the latter approximation derives from $\Delta T \to 0$. The area-normalized resistance becomes:

$$R_b^{''} = R_b \times \text{`area'} \approx \left[\sum_p \frac{1}{2\pi} \int_0^\infty \hbar\omega M_{dD}(\omega)\mathcal{T}(\omega)\frac{\partial f^o_{BE}}{\partial T} d\omega\right]^{-1}. \tag{5.33}$$

The challenge is therefore to evaluate the transmission function $\mathcal{T}(\omega)$, as developed in the following subsections for smooth (acoustic mismatch) and rough (diffuse mismatch) interfaces.

### 5.6.1   *Acoustic Mismatch*

We begin the study of interfaces with the continuum version the 1D atomic chain. Consider a two-segment string that is stretched under a fixed tension $T_e$ as shown in Fig. 5.11. A wavefront of arbitrary displacement form $f_1$ is incident rightward from string 1 on the interface, and the wave then partially reflects ($g$ displacement) and transmits into string 2 ($f_2$ displacement). Basic acoustic theory (French, 1971) reveals that the acoustic velocity in a string can be expressed as:

$$v_a = \sqrt{\frac{T_e}{\mu}}, \tag{5.34}$$

where $\mu$ is the mass density of the string (mass per unit length).

The transverse displacements ($y_1$ and $y_2$) in each string can be expressed as:

$$y_1(x,t) = f_1\left(t - \frac{x}{v_1}\right) + g\left(t + \frac{x}{v_1}\right),$$

$$y_2(x,t) = f_2\left(t - \frac{x}{v_2}\right), \tag{5.35}$$

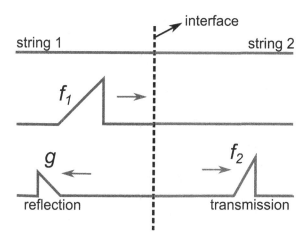

Fig. 5.11   Reflection and transmission of a wave on strings under tension. The wave is partially reflected and transmitted at the interface, where a discontinuity in mass density $\mu$ exists.

with boundary conditions:

$$y_1(0, t) = y_2(0, t), \qquad (5.36)$$

$$\frac{\partial y_1}{\partial x}(0, t) = \frac{\partial y_2}{\partial x}(0, t). \qquad (5.37)$$

Using the interface conditions ($x = 0$) to solve for $f_2$ and $g$ in terms of $f_1$ reveals:

$$f_2(t) = \frac{2v_2}{v_2 + v_1} f_1(t) \qquad (5.38)$$

$$g(t) = \frac{v_2 - v_1}{v_2 + v_1} f_1(t) \qquad (5.39)$$

From the foregoing relations, we can solve for the reflected displacement in terms of the transmitted displacement:

$$g(t) = \frac{v_2 - v_1}{2v_2} f_2(t). \qquad (5.40)$$

These relations lead to the following intuitive observations:

• If $v_1 = v_2$, then nothing is reflected (all transmitted)
• If $v_2 = 0$ (infinite mass), then all is reflected

To this point, we have considered displacements $(y, f, g)$ and velocities $(v_a)$, but our prime focus is energy, specifically the rate of energy flow (French, 1971):

$$P = \frac{1}{2}\mu y_{\text{max}}^2 v_a \omega^2, \tag{5.41}$$

where $y_{\text{max}}$ is the peak displacement, and $\omega$ is the frequency of oscillation.

The ratio of power reflected at the interface to that incident becomes:

$$\frac{P_g}{P_{f1}} = \left(\frac{g}{f_1}\right)^2 = \left(\frac{v_2 - v_1}{v_2 + v_1}\right)^2. \tag{5.42}$$

Instead of velocities, the concept of *acoustic impedance* $(Z)$ is often used:

$$Z = \frac{T_e}{v_a} = \sqrt{T_e \mu} = \mu v_a. \tag{5.43}$$

Then, the normal-direction interfacial energy transmittance from string 1 to string 2 $(t_{12})$ becomes:

$$t_{12} = 1 - \left(\frac{g}{f_1}\right)^2 = 1 - \left(\frac{Z_1 - Z_2}{Z_1 + Z_2}\right)^2 = \frac{4Z_1 Z_2}{(Z_1 + Z_2)^2}. \tag{5.44}$$

Expressed in terms of velocity, the transmittance is:

$$t_{12} = \frac{4v_1 v_2}{(v_1 + v_2)^2}. \tag{5.45}$$

Note that the reverse transmittance $t_{21}$ from string 2 to string 1 is mathematically identical to $t_{12}$ due to the symmetry of the result. This model is called the Acoustic Mismatch Model (AMM) (Little, 1959).

We turn our attention back to the contact-device-contact arrangement to understand the effects of an internal interface within the device as shown in Fig. 5.12. Our prior derivations presumed a uniform number of modes $(M)$ in the device:

$$Q_{\text{ph}} = \frac{1}{2\pi} \int_0^\infty M(\omega) \mathcal{T}(\omega) \hbar \omega \left[f_{BE}^o(T_1) - f_{BE}^o(T_2)\right] d\omega. \tag{5.46}$$

What happens if $M$ changes from one side of the device to the other? For such situations, we must use the concept of interfacial energy transmittance that is specific to a given direction (e.g., $t_{12}$ in Eq. (5.44)).

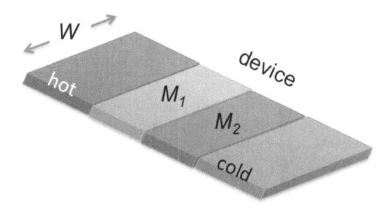

Fig. 5.12  Schematic of a contact-device-contact arrangement in which the number of modes changes at an interface within the device.

The form of Eq. (5.54) that allows for mode discontinuity and direction-specific transmission is:

$$Q_{\text{ph}} = \frac{1}{2\pi} \int_0^\infty \hbar\omega \left[ M_1(\omega)t_{12}(\omega)f_{BE}^o(T_1) - M_2(\omega)t_{21}(\omega)f_{BE}^o(T_2) \right] d\omega.$$
$$(5.47)$$

As a reminder, the number of modes is:

$$M_i(\omega) = \text{'area'} \times \pi \langle v_{g,x} \rangle D_{dD}(\omega), \qquad (5.48)$$

where $i$ denotes the side of the device. The *principle of detailed balance* requires that the integral in Eq. (5.47) must be zero when the two contact temperatures are the same, $T_1 = T_2$. Consequently, the number of modes and transmittances must be related by:

$$M_1(\omega)t_{12}(\omega) = M_2(\omega)t_{21}(\omega) = M_i(\omega)T_i(\omega). \qquad (5.49)$$

This result indicates that the number of modes and transmittance should be thought of as a collective entity for problems in which the number of modes changes across an interface. Moreover, the transmittance itself can depend on the direction of the carrier, in which case the spatial averaging used previously to define the number of modes must be revisited (see Eqs. (4.3)–(4.6) and Eqs. (4.23)–(4.25)).

Extension of the AMM beyond the 1D analysis above provides an example for the process of directional averaging. The multi-dimensional version of the AMM is Little (1959):

$$t_{12}(\theta_1, \omega) = t_{21}(\theta_2, \omega) = \frac{\frac{4Z_2}{Z_1} \cdot \frac{\cos\theta_2}{\cos\theta_1}}{\left(\frac{Z_2}{Z_1} + \frac{\cos\theta_2}{\cos\theta_1}\right)^2}, \qquad (5.50)$$

where $\theta_1$ and $\theta_2$ are the incident and transmitted (refracted) polar angles, as shown in Fig. 5.13. Any frequency dependence in the transmittances would be manifested in the velocity terms that comprise the acoustic impedances and the refracted angle $\theta_2$, although most often the Debye approximation is used in conjunction with the AMM. The incident and transmitted angles are related by Snell's law:

$$\sin\theta_2 = \frac{v_{g2}}{v_{g1}} \sin\theta_1. \qquad (5.51)$$

The directional dependence of $t_{xy}$ necessitates a revisiting of the **k**-space integral expression for heat flow rate. The 3D version can be expressed as:

$$Q_{ph} = \frac{\text{`area'}}{8\pi^3} \int_0^{2\pi} \int_0^{\frac{\pi}{2}} \int_0^{\infty} \hbar\omega \begin{bmatrix} t_{12}(\theta,\omega)v_{g1}\cos\theta\sin\theta f_{BE}^o(T_1) \\ -t_{21}(\theta,\omega)v_{g2}\cos\theta\sin\theta f_{BE}^o(T_2) \end{bmatrix} k^2 dk d\theta d\psi$$

$$= \frac{\text{`area'}\pi}{2\pi} \int_0^{\frac{\pi}{2}} \int_0^{\infty} \hbar\omega \begin{bmatrix} t_{12}(\theta,\omega)v_{g1}\cos\theta\sin\theta D_{1,3D}(\omega) f_{BE}^o(T_1) \\ -t_{21}(\theta,\omega)v_{g2}\cos\theta\sin\theta D_{2,3D}(\omega) f_{BE}^o(T_2) \end{bmatrix} d\omega d\theta, \qquad (5.52)$$

where $D_{i,3D}(\omega)$ is the 3D phonon density of states for side $i$.

The principle of detailed balance requires that the directional ($\theta$) integral over the bracketed term in Eq. (5.52) must be zero when $T_1 = T_2$ (i.e., when $f_{BE}^o(T_1) = f_{BE}^o(T_2)$). Therefore, the following relation must hold:

$$\text{`area'}\pi \left[ \int_0^{\frac{\pi}{2}} t_{12}(\theta,\omega)v_{g1}\cos\theta\sin\theta d\theta \right] D_{1,3D}(\omega)$$

$$= \text{`area'}\pi \left[ \int_0^{\frac{\pi}{2}} t_{21}(\theta,\omega)v_{g2}\cos\theta\sin\theta d\theta \right] D_{2,3D}(\omega)$$

$$\equiv \overline{T}(\omega), \qquad (5.53)$$

where the final equivalence defines a directionally averaged product of the number of modes and transmission function. This function can be used in

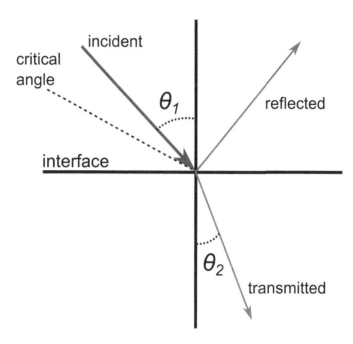

Fig. 5.13   Multi-dimensional reflection and refraction of phonons at an interface under acoustic mismatch. The refraction follows Snell's law.

the general expression for heat flow rate as:

$$Q_{\text{ph}} = \frac{1}{2\pi} \int_0^\infty \overline{\mathcal{T}}(\omega) \hbar\omega \left[ f_{BE}^o(T_1) - f_{BE}^o(T_2) \right] d\omega$$

$$Q_{\text{ph}} = \frac{1}{2\pi} \int_0^\infty M_i(\omega) \mathcal{T}_{AMM,i}(\omega) \hbar\omega \left[ f_{BE}^o(T_1) - f_{BE}^o(T_2) \right] d\omega, \quad (5.54)$$

where the last equality involves a definition of an effective transmission function for the AMM:

$$\mathcal{T}_{AMM,i}(\omega) \equiv \frac{\overline{\mathcal{T}}(\omega)}{M_i(\omega)}. \quad (5.55)$$

Here, the denominator is the previously defined 'number of modes' (see Eq. (5.48)).

The foregoing development reveals a subtle but important observation—namely, that the density of states of only *one* side of the interface needs to be known in order to solve for the overall transport rate. This finding is also observable from Eq. (5.49) for problems with a difference in the number of modes but directionally independent transmittance and is a general consequence of the principle of detailed balance.

The actual function $\overline{T}(\omega)$ for the AMM and other models is complicated because it involves a combination of the directional dependence of Eq. (5.50) and Snell's law (Eq. (5.51)). Little (1959) and Cheeke (1976) used a related functional:[4]

$$\Gamma(\omega) = \int_0^{\frac{\pi}{2}} t_{12}(\theta, \omega) \cos \theta \sin \theta d\theta = \frac{\overline{T}(\omega)}{\text{'area'} \pi D_{1,3D} v_{g1}}$$

$$= \frac{\overline{T}(\omega)}{2M_1} = \frac{1}{2} \mathcal{T}_{AMM,1}(\omega), \tag{5.56}$$

where the frequency dependence of the transmission is retained for generality. We note that any frequency dependence of the foregoing expression would be contained in the transmittance ($t_{12}$). However, under the Debye approximation, the velocities are assumed constant, and this frequency dependence disappears. Tabulated values of $\Gamma$ for the AMM under the Debye approximation have been provided by Cheeke *et al.* (1976) for parameterized acoustic impedances. Instead of a plot of $\Gamma$, we include here a contour graph of the average AMM transmission function $\mathcal{T}_{AMM}$ in Fig. 5.14.

### 5.6.2 *Diffuse Mismatch*

A fundamentally different but also commonly used interface transmission theory is called the Diffuse Mismatch Model (DMM). The term 'diffuse' implies randomness (or something 'spread out'), and in diffuse interface scattering a phonon loses the memory of its origin and its type (branch). In this sense, the DMM can be considered the opposite extreme of the AMM, which generally presumes a retention of phonon coherence. This model applies particularly well to interfaces that are rough in comparison to the carrier wavelength (cf., Fig. 5.2). A carrier moving away from a diffuse interface 'forgets' its original location and branch, and as a result, the transmittance into a particular side is equivalent to reflectance from that side back into itself:

$$t_{12} = r_{21} = 1 - t_{21}, \tag{5.57}$$

---

[4]Note that their term $\alpha_1(\theta)$ is equivalent to $t_{12}(\theta)$ used here.

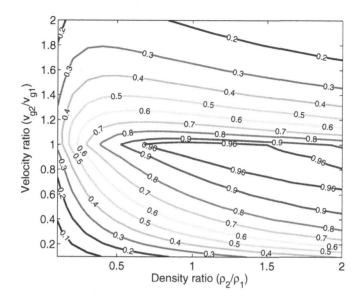

Fig. 5.14  Average AMM transmission coefficient for different group velocities and density ratios.

where the second equality derives from an energy balance (a carrier must be either transmitted or reflected–nothing is absorbed by assumption). The transmittances and reflectances are shown schematically in Fig. 5.15.

Because a scattered phonon can proceed, by assumption, into any branch that is active at its frequency (energy), the summation over branches[5] must be included in the statement of detailed balance that is applicable to the DMM:

$$\sum_p M_1(\omega) t_{12}(\omega) = \sum_p M_2(\omega) t_{21}(\omega) = \sum_p M_2(\omega) \left[1 - t_{12}(\omega)\right]. \quad (5.58)$$

The transmittance does not depend on branch, and therefore, it can be solved as:

$$t_{12}(\omega) = \frac{\sum_p M_2(\omega)}{\sum_p M_1(\omega) + \sum_p M_2(\omega)}. \quad (5.59)$$

---

[5]We have avoided including the branch summations where it is not essential in prior derivations, for the sake of clarity.

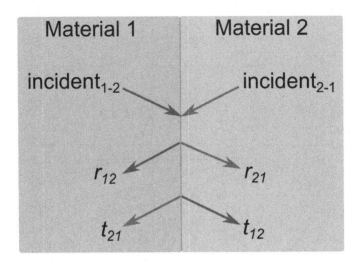

Fig. 5.15 Reflection and transmission at a material interface. If the process is diffuse, then $t_{ij} = r_{ji}$.

The foregoing expression gives the important result that the DMM transmittance is proportional to the fraction of total modes (sides 1 and 2) available on the opposite side of the interface.

$$M_1(\omega)T_1(\omega) = M_1(\omega)\frac{\sum_p M_2(\omega)}{\sum_p M_1(\omega) + \sum_p M_2(\omega)}. \tag{5.60}$$

## 5.7 Thermionic Electron Emission

As stated previously, electron transport is also critically important to heat conduction in bulk metals. Further, coupled electrical-thermal transport is dispositive in determining the performance of many technologically important devices, such as solid-state transistors and thermoelectric materials. Much of the foundational theory associated with the transport of heat by electrons can be obtained from other books in this series by Datta (2012), and Lundstrom and Jeong (2013), particularly when viewed through analogies such as the Wiedemann-Franz law. Here, we focus on the topic of thermionic electron emission through the perspective of modes and transmission developed above. Thermionic transport, while perhaps less common than the ubiquitous transistor, plays a critical role in applications such as electron sources for imaging instruments (e.g., scanning electron microscopes) and in both vacuum (Hatsopoulos and Gyftopoulos, 1973) and solid-state (Shakouri and Bowers, 1997) thermal-to-electrical energy conversion processes.

We begin with a general description of electron emission processes. Figure 5.16 illustrates two possible electron emission pathways from a solid metal into vacuum. Electrons can emit over potential barriers (thermionic emission), or they can tunnel through them (field emission). Field emission involves quantum tunneling through a triangular potential barrier formed by the application of an electric field between the two electrodes (i.e., cathode and anode). Fowler and Nordheim (1928) developed the first theory

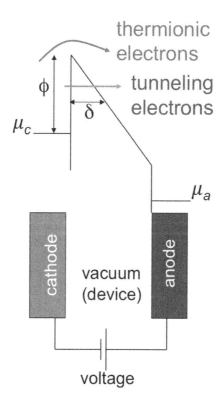

Fig. 5.16   Schematic of thermionic and field electron emission. The power supply creates an electric field through which field-emitted electrons tunnel. Thermionic electrons emit over the energy barrier entirely.

for field emission, for which the tunneling (transmission) probability is proportional to $e^{-\delta(E)}$, where $\delta(E)$ is the local thickness of the potential at a particular energy $E$. Here, we will focus on the process of thermionic emission, which applies to electrons with energies above the surface potential barrier, known as the work function $\phi$.

The general mode-transmission form of the electronic heat flux (cf., Eq. (4.12)) is:

$$J_{Q,\text{el}} = \frac{1}{\pi \hbar} \int_0^\infty M_{dD}(E)(E - \mu)\mathcal{T}(E) \left[ f_{FD}^o(T_1) - f_{FD}^o(T_2) \right] dE, \quad (5.61)$$

where, now, the mode density $M_{dD}$ corresponds to electrons. The primary interest here is to derive the electrical current flux (usually termed 'current density'), which can be obtained by replacing the energy term $(E - \mu)$ in Eq. (5.61) with the elementary electron charge $q$:

$$J = \frac{q}{\pi \hbar} \int_0^\infty M_{dD}(E)\mathcal{T}(E) \left[ f_{FD}^o(T_1) - f_{FD}^o(T_2) \right] dE. \quad (5.62)$$

Thermionic current, by definition, is driven by thermal energy, and the corresponding schematic is shown in Fig. 5.17. We will assume for brevity that the right side reservoir is so cold that no electrons emit from right to left. Consequently, the thermionic current density can be approximated as:

$$J \approx \frac{q}{\pi \hbar} \int_0^\infty M_{dD}(E)\mathcal{T}(E) f_{FD}^o(T_1) dE. \quad (5.63)$$

One complication of thermionic analysis involves the directional dependence of the transmittance, similar to the analysis for phonon interface transmission with acoustic mismatch theory. Consequently, the analysis

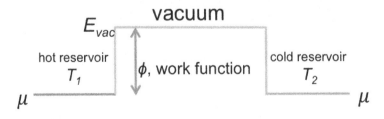

Fig. 5.17  Schematic of a thermionic contact-device-contact arrangement. Here, the device is vacuum, and the same potential barrier exists at each contact interface.

must begin with **k**-space integrals, as did the energy transport analysis (e.g., see Eqs. (4.5) and (4.6) for 2D and 3D, respectively). The general current density in 2D can be expressed as:

$$J = 2q \int_{-\pi/2}^{\pi/2} \int_0^\infty \frac{v_g \cos\theta}{4\pi^2} t_{12}(k,\theta) f_{FD}^o(T_1) k dk d\theta, \qquad (5.64)$$

where $t_{12}$ is the directionally dependent transmittance. Before proceeding further with thermionic emission, we first analyze the 'ideal' result for $t_{12} = 1$:

$$J_{\text{ideal}} = \frac{q}{\pi^2} \int_0^\infty v_g f_{FD}^o(T_1) k dk = \frac{q}{2\pi} \int_0^\infty \left[\frac{2}{\pi} v_g\right] f_{FD}^o(T_1) k dk$$

$$= \frac{q m_e}{2\pi\hbar^2} \int_0^\infty \langle v_{gx} \rangle f_{FD}^o(T_1) dE$$

$$= \frac{q}{\pi\hbar} \int_0^\infty M_{2D}(E) f_{FD}^o(T_1) dE, \qquad (5.65)$$

where the following expressions for 2D density of states and mode density from Chapter 4 have been used with unity valley degeneracy ($g_v = 1$) and the conduction band edge as the zero energy datum ($E_c = 0$):

$$D_{2D}(E) = \frac{m_e}{\pi\hbar^2}, \qquad (5.66)$$

$$M_{2D}(E) = \frac{\pi\hbar}{2} \langle v_{gx} \rangle D_{2D}(E). \qquad (5.67)$$

Equation (5.65) represents the total flux of electrons approaching the potential barrier and is related to the 'supply function' that has been used historically in treatments of thermionic theory (Young, 1959).

The directionality of thermionic transmission derives from the nature of the work function $\phi$, which is an energy barrier that is perpendicular to the the material's surface. Consequently, an electron must possess sufficient energy *associated with the surface normal direction* in order to emit. Assuming as usual a parabolic free-electron energy band, the transmittance is a simple step function that depends on the magnitude of the wavevector ($k$) and its angle $\theta$ from the surface normal:[6]

$$t_{12}(k,\theta) = H\left[\frac{\hbar^2 k^2}{2m_e}\cos^2\theta - (\mu + \phi)\right], \qquad (5.68)$$

---

[6]Note that this transmission function neglects quantum wave effects. The semiclassical models here for thermionic emission are generally quite accurate as compared to exact quantum models (Jensen *et al.*, 2002).

where $H$ is the Heaviside function. Letting $x \equiv \cos^2 \theta$, the expression becomes:

$$J = 2q \int_1^0 \int_0^\infty \frac{-v_g}{4\pi^2 \sqrt{1-x}} H \left[ \frac{\hbar^2 k^2}{2m_e} x - (\mu + \phi) \right] f_{FD}^o(T_1) k dk dx, \quad (5.69)$$

where the factor of 2 and choice of integration limits derive from the symmetry of Eq. (5.68) with respect to $\theta$. The Heaviside function can be used to redefine the integration limits, and the (-) sign reverses the direction of integration in $x$:

$$J = \frac{q}{2\pi^2} \int_{\frac{\mu+\phi}{E(k)}}^1 \int_0^\infty \frac{v_g}{\sqrt{1-x}} H \left[ E(k) - (\mu + \phi) \right] f_{FD}^o(T_1) k dk dx. \quad (5.70)$$

The foregoing integrand essentially requires that $E(k) > (\mu + \phi)$ through the Heaviside function, and if this condition is satisfied, the integral is performed for values of $\cos^2 \theta$ between $(\mu+\phi)/E(k)$ and unity, corresponding to conditions of sufficient energy associated with the surface normal direction for emission to occur.

The following integral identity enables analytic evaluation of the current density:

$$\int_a^b \frac{dx}{\sqrt{1-x}} = -2\sqrt{1-x} \Big|_a^b . \quad (5.71)$$

The current density becomes:

$$J = \frac{q}{\pi^2} \int_0^\infty v_g \sqrt{1 - \frac{E_{\text{vac}}}{E(k)}} H \left[ E(k) - E_{\text{vac}} \right] f_{FD}^o(T_1) k dk, \quad (5.72)$$

where the term $E_{\text{vac}} = \mu + \phi$ represents the vacuum energy level and is used hereafter for brevity.

Comparison of Eq. (5.72) with the ideal case of Eq. (5.65) reveals that the transmission function for this 2D problem must be:

$$T_{2D}(E) = \sqrt{1 - \frac{E_{\text{vac}}}{E}} H \left[ E - E_{\text{vac}} \right] . \quad (5.73)$$

The mode density form of the 2D thermionic current density becomes:

$$J = \frac{q}{\pi \hbar} \int_0^\infty M_{2D}(E) T_{2D}(E) f_{FD}^o(T_1) dE$$

$$= \frac{q}{\pi \hbar} \int_0^\infty M_{2D}(E) \sqrt{1 - \frac{E_{\text{vac}}}{E}} H\left[E - (E_{\text{vac}})\right] f_{FD}^o(T_1) dE$$

$$= \frac{q}{\pi \hbar} \int_{E_{\text{vac}}}^\infty M_{2D}(E) \sqrt{1 - \frac{E_{\text{vac}}}{E}} f_{FD}^o(T_1) dE. \tag{5.74}$$

The foregoing integral is generally not amenable to analytic evaluation; however, by recognizing that the work function $\phi$ is typically much larger than the thermal energy $k_B T$, the distribution function can be approximated as $f_{FD}^o \approx \exp(-(E - \mu)/k_B T)$, which is the Maxwell-Boltzmann distribution. The thermionic current density then becomes:

$$J = \frac{q\sqrt{2 m_e}}{(\pi \hbar)^2} (k_B T)^{3/2} e^{\frac{-\phi}{k_B T}} \int_0^\infty \sqrt{y} e^{-y} dy, \tag{5.75}$$

where $y = (E - E_{\text{vac}})/(k_B T)$. The integral evaluates exactly as $\sqrt{\pi}/2$. Consequently, the expression for 2D thermionic current density becomes:

$$J = \frac{q}{\hbar^2} \sqrt{\frac{m_e}{2}} \left(\frac{k_B T}{\pi}\right)^{3/2} e^{\frac{-\phi}{k_B T}}$$

$$= \left[0.090 \frac{\text{A}}{\text{mK}^{3/2}}\right] T^{3/2} e^{\frac{-\phi}{k_B T}}. \tag{5.76}$$

Similar analyses for 1D and 3D emitters produces the following, general result for the thermionic transmission function:

$$T(E) = \left(1 - \frac{E_{\text{vac}}}{E}\right)^{\frac{d-1}{2}} H\left[E - E_{\text{vac}}\right], \tag{5.77}$$

where $d$ is the problem dimensionality. The resulting general integral for thermionic current density is:

$$\boxed{J = \frac{q}{\pi \hbar} \int_0^\infty M_{dD}(E) \left(1 - \frac{E_{\text{vac}}}{E}\right)^{\frac{d-1}{2}} H\left[E - E_{\text{vac}}\right] f_{FD}^o(T_1) dE.} \tag{5.78}$$

For a 3D emitter, and using the Maxwell-Boltzmann approximation for the distribution function, the result is the well-known Richardson-Dushman equation (Murphy and Good, 1956):

$$J = \int_0^\infty \frac{m_e q}{2\pi^2 \hbar^3} E \frac{\left(1 - \frac{E_{\text{vac}}}{E}\right) H\left[E - E_{\text{vac}}\right]}{1 + \exp\left(\frac{E - \mu}{k_B T_1}\right)} dE$$

$$\approx A T_1^2 \exp\left(\frac{-\phi}{k_B T_1}\right), \tag{5.79}$$

$$\text{where} \quad A = \frac{m_e q k_B^2}{2\pi^2 \hbar^3} = 120 \, \text{A/cm}^2\text{K}^2 \tag{5.80}$$

The transmission functions for the three dimensionalities are shown in Fig. 5.18. The trends reveal that the primary effect of increased dimensionality is to decrease the transmission at a given total energy ($E$) because higher dimensionality provides more degrees of freedom in directions parallel to the surface and its energy barrier. Of course, the 1D case simply produces a step function because the direction of emission is the only possible direction, by definition.

A common method of evaluating an emitter material's work function is to measure its electron energy distribution (EED), which is simply the

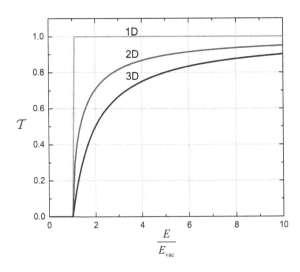

Fig. 5.18    Thermionic transmission as a function of energy for 1D, 2D, and 3D emitters.

spectral current density (the integrand of Eq. (5.78)):

$$\frac{dJ}{dE} = \frac{q}{\pi\hbar}M_{dD}(E)\left(1 - \frac{E_{\text{vac}}}{E}\right)^{\frac{d-1}{2}} H\left[E - E_{\text{vac}}\right] f^o_{FD}(T_1). \qquad (5.81)$$

Figure 5.19 contains a thermionic EED for the (100) face of single-crystal tungsten at 850°C. The 3D EED peaks at an energy that is $k_B T$ above the vacuum energy $E_{\text{vac}}$, while the 2D EED peak occurs at $\frac{1}{2}k_B T$ above the vacuum level. The reason for such a simple outcome is that each degree of freedom not directed normal to the surface will add an average energy of $\frac{1}{2}k_B T$ according to the equipartition theorem (Laurendeau, 2005). For the conditions of Fig. 5.19, the work function for this 3D emitter is 4.56 eV. An EED can also serve to measure the effective temperature of the emitting electrons. For 3D emitters, the full-width half-maximum (FWHM, i.e., the width of the distribution at half of the peak spectral current density) is 2.45 $k_B T$; for 2D emitters, it is 1.80 $k_B T$.

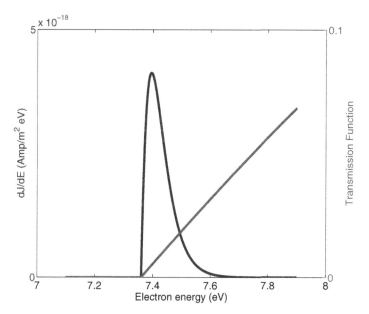

Fig. 5.19 Thermionic electron energy distribution from bulk, single crystal tungsten (100). The data were recorded at an emitter temperature of 850°C.

## 5.8    Conclusion

This final chapter has introduced a broad diversity of important topics necessary to understand thermally driven transport, from a variety of 'internal' scattering mechanisms to interfacial behavior. Each main topic within this chapter has attracted its own cadre of experts, and one could spend an entire career studying any of them. Of course, most learners tend to be generalists, at least initially, and this part of the audience should, having reached the end of the text, be comfortable in progressing deeper into any or all of these interesting areas.

One rather fortuitous outcome of this chapter is the demonstrated cogency and versatility of the mode-transmission Landauer formulation:

- Simple, energy-independent scattering can be easily expressed in transmission form intuitively as the ratio of the mean free path and device length, and at the same time, the ballistic-diffusive transition can be handled through a minor modification to the transmission function.

- More complicated scattering mechanisms, such as three-phonon processes, generally involve parameterized energy (frequency) and temperature dependencies that require numerical solutions to the Landauer integral, but these can be calculated quite readily with contemporary software.

- Interfaces tend to require additional effort because the transmission function becomes directionally dependent for both phonons and electrons. However, the mode-transmission product remains the primary entity to compute.

- Moreover, the treatment of diffuse interfaces is particularly straightforward—the transmission depends simply on the ratio of modes on one side to the total modes on both sides.

- Finally, even thermally driven electrical current—thermionic emission—is intuitively described by the mode-transmission formulation.

This approach seems to differ, at least at first, from more common expositions that tend to employ a variation of the Boltzmann transport equation. However, the final results are the same, while at least conceptually, the mode-transmission formulation offers some advantages in explaining the underlying physics. In the end, the Landauer formulation for real scattering and transmission processes proves to be extremely versatile and effectively unifies the theoretical analysis of these rather complicated and diverse processes. The approach also handles different dimensional spaces in a straightforward and consistent manner. Those who find utility in this approach are encouraged to develop it further as they move past the foundational concepts presented herein to the many exciting contemporary research topics associated with thermal energy at the nanoscale.

**Example Problems**

---

*Problem 5.1: Thermal conductivity of silicon*

Calculate the thermal conductivity of silicon at 300 K using the rudimentary model for transmission given by $\mathcal{T} = \Lambda/(\Lambda + L)$. Assume an energy-independent mean free path of $\Lambda = 200$ nm. Also assume that the three acoustic branches of silicon are replaced by a single branch with a uniform group velocity of 6400 m/s with a Debye temperature of 645 K. Plot the thermal conductivity as a function of length from 10 nm to 20 $\mu$m. Calculate the length beyond which the thermal conductivity changes by less than 2%.

*Solution*

Thermal conductivity $\kappa$ is obtained from the conductance $G_Q$ using the formula $\kappa = G_Q L/\text{'area'}$:

$$\kappa = \frac{L}{\text{'area'}} \frac{1}{2\pi} \int_0^\infty \text{'area'} M_{dD}(\omega)\hbar\omega \frac{\Lambda}{\Lambda+L} \frac{\partial f_{BE}^o}{\partial T} d\omega. \qquad (5.82)$$

In three dimensions and under the Debye approximation, the number of modes is given by:

$$M_{3D}(\omega) = \pi\langle v_g\rangle D(\omega) = \frac{\omega^2}{4\pi v_g^2}. \qquad (5.83)$$

Substituting into Eq. (5.82), we obtain:

$$\kappa = \frac{L\Lambda}{\Lambda+L} \frac{k_B^4 T^3}{8\pi^2 v_g^2 \hbar^3} \int_0^{\theta_D/T} \frac{\chi^4 e^\chi}{(e^\chi - 1)^2} d\chi, \qquad (5.84)$$

where $\chi = \hbar\omega/k_B T$ is the non-dimensional frequency and $\theta_D = \hbar\omega_D/k_B$ is the Debye temperature. Substituting $T = 300$ K, $\Lambda = 200$ nm, $\theta_D = 645$ K, $v_g = 6400$ m/s, we obtain:

$$\kappa = 146.02 \frac{L}{\Lambda+L} \text{ W/m K}. \qquad (5.85)$$

Figure 5.20 shows the length dependence of thermal conductivity. The thermal conductivity is within 2% of its bulk value ($L \to \infty$) beyond a length of approximately 9.5 $\mu$m.

---

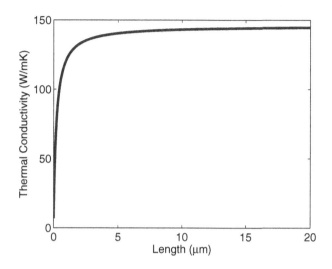

Fig. 5.20 Variation of thermal conductivity as a function of length.

---

*Problem 5.2: Thermal interface resistance*

An experimentalist measures the cryogenic thermal boundary resistance between Si ($\rho_{Si}$ = 2330 kg/m$^3$) and aluminum ($\rho_{Al}$ = 2700 kg/m$^3$) at a temperature of $T$ = 10 K to be $R_b''$ = 4.5 ± 0.3 mm$^2$K/W. We wish to determine the type of interface model that most accurately describes this behavior. To simplify the analysis, first neglect all effects of electronic transport. Also, assume that the longitudinal phonon mode dominates the transport (i.e., neglect all effects of transverse modes). The longitudinal phonon group velocities in silicon and aluminum are: $v_{g,Si}$ = 6400 m/s and $v_{g,Al}$ = 5600 m/s. With these assumptions, answer the following:

(a) Determine the transmission function $\mathcal{T}_{AMM}$ using the acoustic mismatch model by averaging $t_{12}$ over all possible incidence angles. Here the subscript 1 represents Si and 2 represents Al. Hint: The contour plot in Figure 5.21 shows the angular integral $2 \int_0^{\pi/2} t_{12} \cos\theta \sin\theta d\theta$ for different values of density and velocity ratio.

(b) Determine the transmission function $T_{DMM}$ from the diffuse mismatch model.

(c) Determine the thermal boundary resistance using the transmission coeffcients calculated from AMM and DMM. Which model best matches the experimental data?

*Solution*

(a) The transmittance $t_{12}$ from AMM is given by:

$$t_{12} = \frac{4\frac{Z_2}{Z_1}\frac{\cos\theta_2}{\cos\theta_1}}{\left(\frac{Z_2}{Z_1} + \frac{\cos\theta_2}{\cos\theta_1}\right)^2}, \qquad (5.86)$$

where $Z_1$ and $Z_2$ are the acoustic impedances of silicon and aluminum respectively. $\theta_1$ and $\theta_2$ are the angles of incidence (Si side) and refraction (Al side) respectively. $\theta_1$ and $\theta_2$ are related through Snell's law of refraction:

$$\frac{\sin\theta_1}{\sin\theta_2} = \frac{v_{g,1}}{v_{g,2}}. \qquad (5.87)$$

The overall transmission function is an average of Eq. (5.86) over all possible angles of incidence (0 to $\pi/2$). Thus the average transmission coefficient $T_{AMM}$ is given by:

$$T_{AMM} = \frac{\overline{T}}{M} = 2\int_0^{\pi/2} t_{12}\sin\theta_1\cos\theta_1 d\theta_1, \qquad (5.88)$$

where the factor $\sin\theta_1$ in the integrand comes from the integration over all solid angles, and the factor $\cos\theta_1$ comes from the normal component of group velocity, i.e., only the component of group velocity normal to the interface contributes to energy transfer across the interface. Figure 5.21 shows the average transmission function (obtained by evaluating the integral in Eq. (5.88) numerically) for different values of group velocity and density ratios. For $v_{g,2}/v_{g,1} = 5600/6400 \approx 0.9$ and $\rho_2/\rho_1 = 2700/2330 \approx 1.2$, the average transmission function $T_{AMM} = 0.96$.

(b) The transmission function from the diffuse mismatch model $T_{DMM}$ is given by:

$$T_{DMM} = \frac{1/v_{g,2}^2}{1/v_{g,1}^2 + 1/v_{g,2}^2}. \qquad (5.89)$$

Note that the above expression is valid only under the Debye approximation. Substituting $v_{g,1} = 6400$ m/s and $v_{g,2} = 5600$ m/s, we obtain $T_{DMM} = 0.56$.

(c) The area normalized interface resistance $R_b''$ is given by:

$$R_b'' = \left[ \frac{1}{2\pi} \int_0^\infty \hbar\omega M_{3D}(\omega) T(\omega) \frac{\partial f_{BE}^o}{\partial T} d\omega \right]^{-1}, \qquad (5.90)$$

where we have dropped the summation over phonon polarization since we are considering only the longitudinal branch in this problem. Making a change of variable given by $\chi = \hbar\omega/k_B T$, we obtain:

$$R_b'' = \left[ \frac{k_B^4 T^3 T}{8\pi^2 v_{g,Si}^2 \hbar^3} \int_0^{\theta_{D,Si}/T} \frac{\chi^4 e^\chi}{(e^\chi - 1)^2} d\chi \right]^{-1}. \qquad (5.91)$$

For small temperatures in comparison to the Debye temperature, the upper limit of the integral can be approximated as infinity, and the integral evaluates to $4\pi^4/15$. Thus for low temperatures, the thermal boundary resistance $R_b''$ is given by:

$$R_b'' = \left[ \frac{\pi^2}{30} \frac{k_B^4}{\hbar^3 v_{g,Si}^2} T \right]^{-1} T^{-3}. \qquad (5.92)$$

From the above equation, we obtain the following values for $R_b''$.

$$R_b''(T_{AMM}, 10 \text{ K}) = 4.18 \times 10^{-6} \text{ m}^2\text{K/W} \qquad (5.93)$$
$$= 4.18 \text{ mm}^2\text{K/W} \qquad (5.94)$$
$$R_b''(T_{DMM}, 10 \text{ K}) = 7.18 \times 10^{-6} \text{ m}^2\text{K/W} \qquad (5.95)$$
$$= 7.18 \text{ mm}^2\text{K/W} \qquad (5.96)$$

From the above values, the AMM prediction is closer to experimental results.

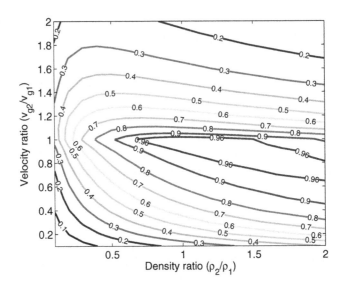

Fig. 5.21   Angular average of the transmission coefficient from AMM for different group velocities and density ratios.

## Problem 5.3: Thermionic emission

An electron emission material has a chemical potential $\mu = 5$ eV and a work function $\phi = 2.36$ eV.

(a) Calculate the current flux at a cathode temperature of 400 K.
(b) Calculate the electron energy at which the thermionic electron energy distribution is a maximum.
(c) Make a qualitative plot of the transmission function and the electron energy distribution as a function of energy.

*Solution*

(a) The current flux is obtained from the Richardson-Dushman equation:

$$J_{3D} = \frac{mk_B^2 q}{2\pi^2\hbar^3}T_1^2 \exp\frac{-\phi}{k_BT_1} = AT_1^2 \exp\frac{-\phi}{k_BT_1}. \qquad (5.97)$$

Substituting $T_1 = 400$ K and $\phi = 2.36$ eV, $A = 1.2 \times 10^6$ A/m$^2$K$^2$, we obtain $J_{3D} = 3.87 \times 10^{-19}$ A/m$^2$.

(b) The maximum in the thermionic electron energy distribution (TEED) occurs at an energy of $E_{max} = \mu + \phi + k_B T_1 = 7.39$ eV.

(c) Figure 5.22 shows a plot of TEED and the transmission function as a function of electron energy.

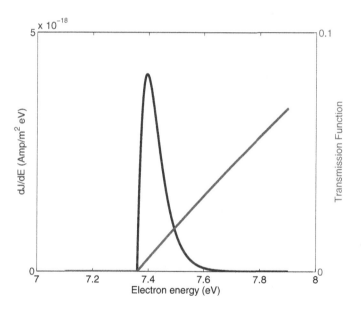

Fig. 5.22 Thermionic electron energy distribution and transmission function.

**Problem 5.4: Thermal conductivity of confined nanostructures**

Confining the dimensions of a bulk material via nanostructuring can produce a drastic reduction in the thermal conductivity. Sketches of three silicon nanostructures are drawn in Figure 5.23 (not to scale). The confined dimensions are indicated. Assume the average group velocity of LA and TA modes of silicon as $v_{g,Si} = 5200$ m/s. The Debye approximation can be invoked for these acoustic branches. Dominant scattering phenomena that occur in such nanostructures include temperature independent boundary scattering $\tau_b^{-1}$, Rayleigh-type (Klemens model) impurity scattering $\tau_i^{-1}$, and temperature dependent Umklapp scattering $\tau_{U,p-p}^{-1}$:

$$\tau_b^{-1} = \frac{v_g}{\sqrt{l_1 l_2}}, \quad \tau_i^{-1} = A\omega^4, \quad \tau_{U,p-p}^{-1} = BT\omega^2 e^{-C/T}, \qquad (5.98)$$

where the constants $A$, $B$ and $C$ are given in Table 5.1.

(a) Calculate the thermal conductivity of the three nanostructures shown in Figure 5.23 at $T = 300$ K. Include the aforementioned scattering models. You may need to use a numerical solver (e.g., WolframAlpha) to evaluate the integral.

(b) The online Chapter 5 CDF[7] tool plots thermal conductivity of silicon and germanium nanostructures as a function of temperature. The model incorporates the scattering phenomenon discussed in part (a) of the problem. The CDF allows the user to choose the material, Si or Ge. The tool also allows the user to vary the boundary size from 10 nm to 200 nm. This range is chosen for accurate applicability of the above scattering phenomenon. Observe the change in thermal conductivity as the boundary size changes. Provide a physical explanation for the above trend. Also, assess the temperature dependence of thermal conductivity.

*Solution*

(a) Thermal conductivity is given by:

$$\kappa = \frac{1}{2\pi} \int_0^\infty \Lambda(\omega) M_{3D}(\omega) \hbar \omega \frac{\partial f_{BE}^o}{\partial T} d\omega$$

$$= \frac{1}{2\pi} \int_0^{\omega_D} \frac{v_g}{\tau^{-1}(\omega)} \pi \frac{v_g}{2} \frac{\omega^2}{2\pi^2 v_g^3} \hbar \omega \frac{\partial f_{BE}^o}{\partial T} d\omega, \qquad (5.99)$$

---

[7]See http://nanohub.org/groups/cdf_tools_thermal_energy_course/wiki

where the effective scattering rate $\tau^{-1}(\omega)$ is given by Matthiessen's rule:

$$\tau^{-1}(\omega) = \tau_i^{-1}(\omega) + \tau_b^{-1}(\omega) + \tau_{U,p-p}^{-1}(\omega). \tag{5.100}$$

The integral in Eq. (5.99) needs to be evaluated numerically for different strengths of the boundary scattering term. Performing this integration, we obtain thermal conductivities of 3.5, 12.2 and 24.6 W/mK for the first, second and third nanostructures respectively. As expected, the thermal conductivity increases with size of the sample as the strength of boundary scattering varies inversely with the size.

(b) The online Chapter 5 CDF tool can be used to observe the temperature dependence of thermal conductivity for Si and Ge nanostructures. The tool can also be used to observe the change in thermal conductivity with variation in the size of the sample. Figure 5.24 shows a snapshot from the CDF tool and reinforces the idea that thermal conductivity increases with size of the nanostructure. The graphs also show a $T^3$ dependence of thermal conductivity at low temperatures (similar to specific heat) and the reduction in thermal conductivity at high temperatures is due to the dominance of Umklapp scattering.

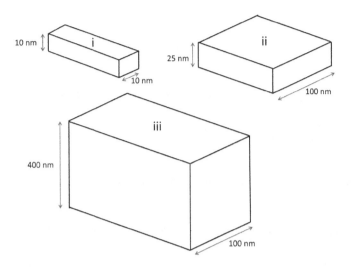

Fig. 5.23　Schematic of confined nanostructures.

Table 5.1　Scattering parameters for silicon and germanium

| Material | A $(s^3)$ | B $(s/K)$ | C (K) | $\theta_D$ (K) |
|---|---|---|---|---|
| silicon | $1.32 \times 10^{-45}$ | $1.73 \times 10^{-19}$ | 137.39 | 452.8 |
| germanium | $2.4 \times 10^{-44}$ | $8.8 \times 10^{-20}$ | 57.6 | 202.7 |

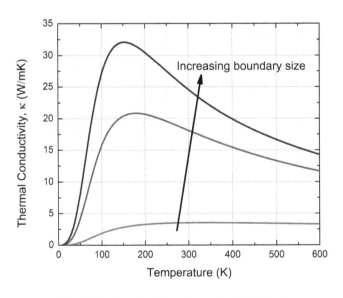

Fig. 5.24　Snapshot from the online CDF tool.

## Appendix A

# The Graphene ZA Branch

## A.1 Introduction

The ZA phonon branch of graphene, shown in Fig. A.1 is quite peculiar among the various types of phonon dispersion relations found in nature. The ZA modes are important because even though their average group velocity is low, they can contribute disproportionately to thermal energy storage and heat conduction (Singh *et al.*, 2011b). Often, the ZA dispersion's quadratic character is presented as common knowledge, as if its form should be obvious. However, new learners often do not possess such intuition, and this appendix's purpose is to provide a basis for the ZA dispersion using continuum concepts.

## A.2 Geometry and Governing Equation

The ZA branch derives from the so-called 'flexural mode' of plate bending, which is an out-of-plane displacement as shown in Fig. A.2. In fact, continuum plate theory suffices for our purposes to derive the characteristic quadratic dispersion. The variable $w$ will represent local displacement in the $z$ direction.

The governing equation for vertical displacement $w$ derives from Hamilton's principle, which is a variational conservation principle that balances the kinetic and potential energies within a deformable body with the applied load. For the case of plate flexural deformation, the governing equation becomes (Doyle, 1997):

$$\mathcal{D}\nabla^2\nabla^2 w + \rho h \frac{\partial^2 w}{\partial t^2} = \mathcal{F}. \tag{A.1}$$

where

$$\mathcal{D} \equiv \frac{E_Y h^3}{12(1 - \nu^2)} = \text{plate bending stiffness}$$

$\rho = \text{plate density}$

$h = \text{plate thickness}$

$\mathcal{F} = \text{plate loading (per area)}$

$E_Y = \text{plate modulus}$

$\nu = \text{Poisson ratio}$

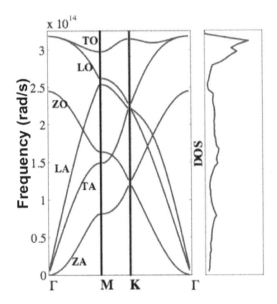

Fig. A.1   Phonon dispersion curves for graphene and the corresponding density of states derived using a modified Tersoff potential (Singh *et al.*, 2011b).

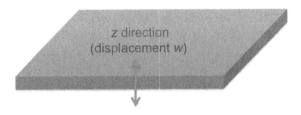

Fig. A.2   Schematic showing the vibrational displacement direction of the flexural mode of a thin plate.

## A.3 Solution and Dispersion

As we did for time-varying displacements of the atomic chain, we can assume a periodic solution for $w(t)$:

$$w(x,t) = \hat{w}(x)e^{i\omega t}, \tag{A.2}$$

where $\hat{w}$ is the amplitude of displacement. Upon substitution and simplification, the homogeneous governing equation for displacement amplitude becomes:

$$\frac{\partial^2 \hat{w}}{\partial x^2} \pm \beta^2 \hat{w} = 0, \tag{A.3}$$

where

$$\beta^2 = \omega \sqrt{\frac{\rho h}{\mathcal{D}}}. \tag{A.4}$$

The veracity of the foregoing result can be inferred from considering the homogeneous form of the original governing equation (Eq. (A.1)) in the frequency domain as:

$$\left(\nabla^2 \nabla^2 - \beta^4\right) \hat{w} = 0, \tag{A.5}$$

for which simple substitution confirms that Eq. (A.3) is an intermediate solution.

The full solution of Eq. (A.3) can be expressed as a plane wave:

$$\hat{w}(x) \sim \exp(-iK_j x), \tag{A.6}$$

where $K_j$ are the allowable wavevectors corresponding to the eigenvalues of Eq. (A.3):

$$K_1 = \pm\sqrt{\omega}\left(\frac{\rho h}{\mathcal{D}}\right)^{1/4}, \tag{A.7}$$

$$K_2 = \pm i\sqrt{\omega}\left(\frac{\rho h}{\mathcal{D}}\right)^{1/4}. \tag{A.8}$$

Rearranging and allowing only positive values of the resolved frequency produces the dispersion relation:

$$\boxed{\omega = \sqrt{\frac{\mathcal{D}}{\rho h}}K^2}. \tag{A.9}$$

This result reveals that flexural plate bending indeed produces a quadratic dispersion, and in effect, the plate bending stiffness $\mathcal{D}$ serves the role of the spring constant from the discrete analysis of the atomic chain. The corresponding group velocity is proportional to $\sqrt{\omega}$:

$$v_g = \frac{d\omega}{dK} = 2\sqrt{\omega}\left(\frac{\mathcal{D}}{\rho h}\right)^{1/4}. \qquad (A.10)$$

The elastic bending stiffness of graphene depends on the crystal orientation and is generally $\mathcal{D} \approx 1$ eV ($= 1.6 \times 10^{-19}$ Nm), and graphene's areal mass density ($\rho h$) is $7.6 \times 10^{-7}$ kg/m$^2$ (Lu *et al.*, 2009). Taking a specific value of $\mathcal{D} = 1.9 \times 10^{-19}$ Nm produces the following dispersion:

$$\omega = \left(5 \times 10^{-7}\frac{m^2}{s}\right)K^2. \qquad (A.11)$$

The resulting ZA branch is shown in Fig. A.3.

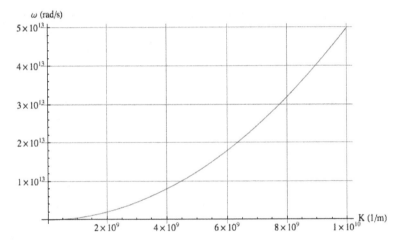

Fig. A.3  Approximation of the ZA branch using a continuum flexural plate bending model.

# Electron and Phonon Contributions to Heat Conduction in Graphene

## B.1 Introduction

Graphene is a material that has elicited tremendous interest in the research and technology communities since the early 2000s because of its unique electrical properties, and to a lesser extent its thermal properties. Graphene's unique electronic band structure (i.e., dispersion) produces high 2D conductivity and has motivated numerous concepts for use in three-terminal devices (Das Sarma, *et al.*, 2011). Unlike most materials, the electronic $E(k)$ dispersion in graphene is linear near the intrinsic Fermi level (or Dirac point), much like photons or long-wavelength acoustic phonons, as shown in Fig. B.1. Even though graphene exhibits outstanding electrical conduction properties, its thermal properties are dominated by phonons, and here we explain this outcome qualitatively using the concept of 2D mode density, $M_{2D}$.

## B.2 Mode Densities

This linear electronic dispersion produces a 2D mode density that is expressed as (Lundstrom and Jeong, 2013):

$$M_{2D}(E) = \frac{2|E|}{\pi \hbar v_F},$$ (B.1)

where $E$ is measured from the Dirac point, and $v_F$ is the constant electron velocity (because of the linear dispersion), $v_F \approx 10^6$ m/s.

For phonons, the 2D mode density can be expressed generally as (see Eq. (4.18)):

$$M_{2D}(\omega) = \frac{K(\omega)}{\pi} = 2v_g(\omega)D_{2D}(\omega).$$ (B.2)

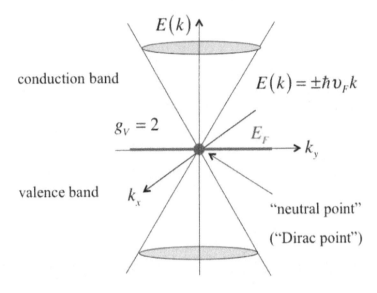

Fig. B.1    Band structure of graphene near the Dirac point. Figure from Lundstrom and Jeong (2013), used with permission.

Using the Debye approximation for the LA and TA phonon branches (see Eqs. (2.50) and (2.51)) and quadratic dispersion for the ZA branch (Eq. (A.11)), the mode densities become:

$$M_{2D,LA}(\omega) = \frac{\omega}{\pi v_{g,LA}}, \tag{B.3}$$

$$M_{2D,TA}(\omega) = \frac{\omega}{\pi v_{g,TA}}, \tag{B.4}$$

$$M_{2D,ZA}(\omega) = \frac{1}{\pi}\sqrt{\frac{\omega}{C}}, \tag{B.5}$$

where $C = 5 \times 10^{-7}$ m$^2$/s. Debye group velocities for the LA and TA branches in graphene are $v_{g,LA} = 2 \times 10^4$ m/s and $v_{g,TA} = 1.5 \times 10^4$ m/s (Singh *et al.*, 2011b).

Notably, the electron mode density of Eq. (B.1) is essentially identical (with a factor of 2 for spin) to that of the linear phonon branches (LA and TA) in Eqs. (B.3) and (B.4). The major quantitative difference is that the electron velocity $v_F$ is two orders of magnitude larger than the phonon group velocities, making the electronic mode density much smaller than those of the linear acoustic branches.

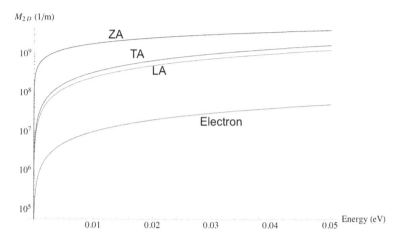

Fig. B.2   2D mode densities of graphene for electrons and the three active phonon branches between 0 and 50 meV. The electronic result uses a linear dispersion approximation, while phonon approximations are linear for the LA and TA branches and quadratic for the ZA branch.

Figure B.2 shows the resulting variation of 2D mode densities for electrons and all acoustic phonon branches over the energy range from zero to 50 meV, which is approximately $2k_BT$ at room temperature. The results indeed show that the electronic mode density is one to two orders of magnitude below those of phonons. Consequently, we expect notionally the phonons to dominate heat conduction. The figure shows further that the ZA mode density is much higher than any others over the plotted energy range. This result suggests that ZA modes will be prevalent thermal carriers, at least up to room temperature.

## B.3   Thermal Conductivity

Of course the mode density does not itself dictate a material's thermal conductivity completely. A generalized version of Eq. (5.7) that encompasses all carriers and branches in graphene as an energy integral is:

$$\kappa_{2D} = \sum_i \frac{1}{2\pi\hbar} \int M_{dD}(E)\Lambda(E)(E - \mu)\frac{\partial f_i^o}{\partial T}dE, \qquad (B.6)$$

where $i$ represents electrons and each phonon branch.

We note that the scattering term, as discussed in Chapter 5, can be very complicated and has elicited its share of scholarly discourse (Seol *et al.*, 2011b; Singh *et al.*, 2011a). Further, the temperature derivative term in Eq. (B.6) differs for electrons and phonons, but as shown in Figs. 4.3 and 4.4, this term's magnitude is similar near the peak in thermal conductance (i.e., near $E \approx k_B T$). Despite the absence of a full quantitative analysis here, the extremely large difference among mode densities between electrons and phonons suggests that phonons should dominate heat conduction, and indeed, experimental measurements corroborate this notion. The total thermal conductivity of substrate-supported graphene has been measured to be of order 100 to 1000 W/mK (using the intergraphene plane spacing of 0.34 nm as the cross-section height) (Seol *et al.*, 2011b), whereas the electron-only contribution has been measured recently to be of order 1 to 10 W/mK (Tayari *et al.*, 2013)—i.e., the electron contribution to thermal conductivity is one to two orders of magnitude less than that of phonons.

# Bibliography

Ashcroft, N. W. and Mermin, N. D. (1976). *Solid State Physics* (Saunders College Publishing, Philadelphia, PA).

Barr, E. S. (1960). Historical Survey of the Early Development of the Infrared Spectral Region, *American Journal of Physics* **28**, 1, p. 42.

Berman, R., Simon, F. E. and Ziman, J. M. (1953). The Thermal Conductivity of Diamond at Low Temperatures, *Proceedings of the Royal Society of London. Series A. Mathematical and Physical Sciences* **220**, 1141, pp. 171–183.

Callaway, J. (1959). Model for Lattice Thermal Conductivity at Low Temperatures, *Physical Review* **113**, 4, p. 1046.

Chandler, D. (1987). *Introduction to Modern Statistical Mechanics*, 1st edn. (Oxford University Press, USA).

Cheeke, J., Ettinger, H. and Hebral, B. (1976). Analysis of Heat Transfer Between Solids at Low Temperatures, *Canadian Journal of Physics* **54**, 17, pp. 1749–1771.

Chen, G. (2005). *Nanoscale Energy Transport and Conversion*, A Parallel Treatment of Electrons, Molecules, Phonons, and Photons (Oxford University Press, USA).

Das Sarma, S., Adam, S., Hwang, E. H. and Rossi, E. (2011). Electronic Transport in Two-dimensional Graphene, *Reviews of Modern Physics* **83**, 2, pp. 407–470.

Datta, S. (2012). *Lessons from Nanoelectronics*, A New Perspective on Transport (World Scientific Publishing Company).

De Graef, M. and McHenry, M. E. (2012). *Structure of Materials*, An Introduction to Crystallography, Diffraction and Symmetry (Cambridge University Press).

Debye, P. (1912). Zur Theorie der Spezifischen Wärmen, *Annalen der Physik* **344**, 14, pp. 789–839.

Doyle, J. F. (1997). *Wave Propagation in Structures: Spectral Analysis Using Fast Discrete Fourier Transforms*, Mechanical Engineering Series (Springer-Verlag).

Einstein, A. (1906). Die Plancksche Theorie der Strahlung und die Theorie der Spezifischen Wärme, *Annalen der Physik* **327**, 1, pp. 180–190.

Fowler, R. H. and Nordheim, L. (1928). Electron Emission in Intense Electric Fields, *Proceedings of the Royal Society of London Series a-Containing Papers of a Mathematical and Physical Character* **119**, pp. 173–181.

French, A. P. (1971). *Vibrations And Waves*, M.I.T. Introductory Physics Series (CRC Press, Boca Raton, FL).

Hatsopoulos, G. N. and Gyftopoulos, E. P. (1973). *Thermionic energy conversion* (MIT, Cambridge, MA).

Holland, M. G. (1963). Analysis of Lattice Thermal Conductivity, *Physical Review* **132**, 6, p. 2461.

Jensen, K., O Shea, P. and Feldman, D. (2002). Generalized Electron Emission Model for Field, Thermal, and Photoemission, *Applied Physics Letters*.

Joshi, A. A. and Majumdar, A. (1993). Transient Ballistic and Diffusive Phonon Heat Transport in Thin Films, *Journal of Applied Physics* **74**, 1, pp. 31–39.

Kapitza, P. L. (1941). The Study of Heat Transfer in Helium II, *Journal of Physics-USSR* **4**, 1-6, pp. 181–210.

Kaviany, M. (2008). *Heat Transfer Physics* (Cambridge University Press).

Kittel (2007). *Introduction to Solid State Physics*, 7th Ed (John Wiley & Sons).

Klemens, P. G. (1951). The Thermal Conductivity of Dielectric Solids at Low Temperatures (Theoretical), *Proceedings of the Royal Society London, Series A* **208**, pp. 108–133.

Klemens, P. G. (1958). Thermal Conductivity and Lattice Vibration Modes, *Solid State Physics-Advances in Research and Applications* **7**, pp. 1–98.

Laurendeau, P. N. M. (2005). *Statistical Thermodynamics: Fundamentals and Applications* (Cambridge University Press).

Lien, W. H. and Phillips, N. E. (1964). Low-Temperature Heat Capacities of Potassium, Rubidium, and Cesium, *Physical Review* **133**, 5A, pp. A1370–A1377.

Little, W. (1959). The Transport of Heat Between Dissimilar Solids at Low Temperatures, *Canadian Journal of Physics*.

Lu, Q., Arroyo, M. and Huang, R. (2009). Elastic Bending Modulus of Monolayer Graphene, *Journal of Physics D: Applied Physics* **42**, 10, p. 102002.

Lundstrom, M. (2009). *Fundamentals of Carrier Transport*, 2nd edn. (Cambridge University Press).

Lundstrom, M. and Jeong, C. (2013). *Near-equilibrium Transport: Fundamentals and Applications (Lessons from Nanoscience: A Lecture Note Series)* (World Scientific Publishing Company).

Modest, M. F. (2003). *Radiative Heat Transfer* (Academic Press).

Murphy, E. L. and Good, R. H. (1956). Thermionic Emission, Field Emission, and the Transition Region, *Physical Review* **102**, 6, pp. 1464–1473.

Ni, C. and Murthy, J. Y. (2012). Phonon Transport Modeling Using Boltzmann Transport Equation With Anisotropic Relaxation Times, *Journal of Heat Transfer* **134**, 8, p. 082401.

Pathria, R. K. and Beale, P. D., Statistical Mechanics (Third Edition) (Academic Press, Boston, 2011), pp. 179–229.

Pop, E., Varshney, V. and Roy, A. K. (2012). Thermal Properties of Graphene: Fundamentals and Applications, *MRS Bulletin* **37**, 12, pp. 1273–1281.

Rego, L. and Kirczenow, G. (1998). Quantized Thermal Conductance of Dielectric Quantum Wires, *Physical Review Letters* **81**, pp. 232–235.

Rybicki, G. B. and Lightman, A. P. (2008). *Radiative Processes in Astrophysics* (Wiley-VCH).

Saha, S. K., Waghmare, U. V., Krishnamurthy, H. R. and Sood, A. K. (2008). Phonons in Few-Layer Graphene and Interplanar Interaction: A First-Principles Study, *Physical Review B* **78**, 16, p. 165421.

Seol, G., Yoon, Y., Fodor, J. K., Guo, J., Matsudaira, A., Kienle, D., Liang, G., Klimeck, G., Lundstrom, M. and Saeed, A. I. (2011a). CNTbands, http://nanohub.org/resources/1838.

Seol, J. H., Moore, A. L., Shi, L., Jo, I. and Yao, Z. (2011b). Thermal Conductivity Measurement of Graphene Exfoliated on Silicon Dioxide, *Journal of Heat Transfer* **133**, 2, p. 022403.

Shakouri, A. and Bowers, J. E. (1997). Heterostructure Integrated Thermionic Coolers, *Applied Physics Letters* **71**, 9, pp. 1234–1236.

Singh, D., Murthy, J. Y. and Fisher, T. S. (2011a). Mechanism of Thermal Conductivity Reduction in Few-layer Graphene, *Journal of Applied Physics* **110**, 4, pp. 044317–044317.

Singh, D., Murthy, J. Y. and Fisher, T. S. (2011b). Spectral Phonon Conduction and Dominant Scattering Pathways in Graphene, *Journal of Applied Physics* **110**, 9, p. 094312.

Tayari, V., Island, J. O., Porter, J. M. and Champagne, A. R. (2013). Electronic Thermal Conductivity Measurements in Intrinsic Graphene, *arXiv.org*.

Tersoff, J. (1988). New Empirical Approach for the Structure and Energy of Covalent Systems, *Physical Review B* **37**, 12, pp. 6991–7000.

Vincenti, W. G. and Kruger, C. H. (1967). *Introduction to Physical Gas Dynamics* (Krieger Publishing Co.).

Wang, X. (2012). *Experimental Micro/Nanoscale Thermal Transport* (Wiley).

Yang, R. and Chen, G. (2004). Thermal Conductivity Modeling of Periodic Two-Dimensional Nanocomposites, *Physical Review B* **69**, 1, p. 195316.

Young, D. A. and Maris, H. J. (1989). Lattice-Dynamical Calculation of the Kapitza Resistance Between FCC Lattices, *Physical Review B* **40**, 6, pp. 3685–3693.

Young, R. D. (1959). Theoretical Total-Energy Distribution of Field-Emitted Electrons, *Physical Review* **113**, 1, pp. 110–114.

Zhang, W., Fisher, T. S. and Mingo, N. (2007). The Atomistic Green's Function Method: An Efficient Simulation Approach for Nanoscale Phonon Transport, *Numerical Heat Transfer, Part B: Fundamentals* **51**, 4, pp. 333–349.

Zhang, Z. (2007). *Nano/Microscale Heat Transfer* (McGraw-Hill Professional).

Ziman, J. (1972). *Principles of the Theory of Solids*, 2nd edn. (Cambridge University Press, Cambridge).

# Index